Thermodynamics of Irreversible Processes

BERNARD H. LAVENDA
Università degli Studi, Camerino

DOVER PUBLICATIONS, INC.
New York

Published in Canada by General Publishing Company, Ltd., 30 Lesmill Road, Don Mills, Toronto, Ontario.
Published in the United Kingdom by Constable and Company, Ltd., 3 The Lanchesters, 162–164 Fulham Palace Road, London W6 9ER.

This Dover edition, first published in 1993, is an unabridged republication of the work first published by The Macmillan Press Ltd., London, in 1978.

Manufactured in the United States of America
Dover Publications, Inc., 31 East 2nd Street, Mineola, N.Y. 11501

Library of Congress Cataloging-in-Publication Data

Lavenda, Bernard H.
Thermodynamics of irreversible processes / Bernard H. Lavenda.
p. cm.
Originally published: London : Macmillan, 1978.
Includes index.
ISBN 0-486-67576-9 (pbk.)
1. Irreversible processes. I. Title.
QD501.L32 1993
536'.7—dc20 93-3190
CIP

To Marlene

Contents

Preface ix

Introduction xi
Formulations of non-equilibrium thermodynamics xi
Quasi-thermodynamic approach to nonlinear thermodynamics xii

1 *Equilibrium Thermodynamics* 1
1.1 Carathéodory's theory 3
1.1.1 Definitions and conditions of equilibrium 3
1.1.2 The first law 4
1.1.3 The second law 5
1.2 Gibbsian thermodynamics 11
1.3 Geometry of the Gibbs space 14
1.4 Equilibrium extremum principles 19
1.5 Principle of maximum work 21

2 *Classical Non-equilibrium Thermodynamics* 25
2.1 The predecessors of Onsager 25
2.2 The statistical basis of Onsager's reciprocal relations 29
2.3 Criticisms of classical non-equilibrium thermodynamics 34

3 *'Rational' Thermodynamics* 41
3.1 Axioms of rational thermodynamics 42
3.1.1 Axiom of admissibility 42
3.1.2 Axiom of determinism 44
3.1.3 Axiom of equipresence 44
3.1.4 Axiom of material frame-indifference or objectivity 44
3.2 The formalism of rational thermodynamics 45
3.3 The interpretations of 'dissipation' 49
3.4 Limitations of rational thermodynamics 55

4 *'Generalised' Thermodynamics* 60
4.1 The development of generalised thermodynamics 62
4.2 Thermodynamic *versus* kinetic stability criteria 69

5 *Nonlinear Thermodynamics* 76
5.1 The thermodynamic principle of the balance of power 79
5.2 The balance equation of mechanical power in the entropy representation 82
5.3 Properties and forms of the balance equation of mechanical power 88

6 *Non-equilibrium Variational Principles* 93
6.1 The dynamic Le Châtelier principle 94
6.2 The principle of least dissipation of energy 97
6.3 Gauss's principle in non-equilibrium thermodynamics 101
6.4 Variational principles of nonlinear thermodynamics 103
6.5 Kinetic formulation of thermodynamic variational principles 105

7 *Quasi-Thermodynamic Stability Theory* 111
7.1 Elements of kinetic stability theory 112
7.2 The complex power method 115
7.3 The significance of the antisymmetric components of the phenomenological coefficient matrices 119
7.4 Stability of non-equilibrium stationary states 120
7.5 The mechanical bases of phenomenological symmetries and antisymmetries 123
7.6 Stability of nonlinear irreversible processes 127

8 *Field Thermodynamics* 132
8.1 The velocity potential analysis of multistationary state transitions 134
8.2 Thermokinematics of rotational non-equilibrium processes 136
8.3 Thermodynamics of force fields 140
8.4 Elements of the field theory 141
8.5 Thermodynamic principles of the field 144
8.6 Description of fields by thermodynamic variational principles 147

9 *Continuum Thermodynamics* 150
9.1 The balance equations of the continua 151
9.2 The generalised power equation 153
9.3 Variational equations of the continua 157
9.4 Thermodynamic variational principles of the continua 161
9.5 An illustration of the internal state variable representation 163
9.6 Thermodynamic evolutionary criteria 167
9.7 Thermodynamics of nonlinear dissipative wavetrains 168

Glossary of Principal Symbols 175

Index 179

Preface

In view of the large amount of recent work on the thermodynamics of non-equilibrium processes, there is a real need for a book giving a clear exposition of the thermodynamic formalism applicable to nonlinear thermodynamic processes at a mathematical level, accessible to physicists and to theoretically inclined biologists and chemists. This book attempts to fill this need by making a definite statement in regard to the present-day status of linear and nonlinear thermodynamics.

For the past two decades, two schools of non-equilibrium thermodynamics have dominated the literature: the school of 'rational' thermodynamics of Coleman and co-workers and the school of 'generalised' thermodynamics that is associated with the names of Glansdorff and Prigogine. Although both these schools praise themselves for their all-embracing coverage of the field, I have never seen 'generalised' thermodynamics applied to elastic materials with memory, nor have I seen 'rational' thermodynamics used in the analysis of chemical instabilities. The apparent incompatibility of the theories may bewilder the reader who wants to understand what non-equilibrium thermodynamics is all about.

Moreover, the vastly different usage of concepts and notations has not helped matters. A case in point is the ambiguity in the meanings of 'dissipation' and 'irreversibility'. These terms are often regarded as synonymous, and the precise meanings of each left in doubt. Confusion also arises over the meaning of the term 'nonlinear' in thermodynamics. While everyone knows what a nonlinear differential equation looks like, its usage in thermodynamics is far from being self-evident.

In this book I attempt to resolve such types of ambiguity and misconception. After the introductory chapter on equilibrium thermodynamics, which serves to form a common background and as a reference for all future developments, the book is divided into two parts. The first part (chapters 1–4) is a critical analysis of the classical theory of non-equilibrium thermodynamics and its more recent offshoots. The second part (chapters 5–9) deals with my own interpretation of what a theory of non-equilibrium thermodynamics should include. In the same way that equilibrium thermodynamics offers criteria for its validity so, too, non-equilibrium thermodynamics must provide for similar criteria. My point of view is that these criteria must ultimately come from nonlinear mechanics and kinetic stability theory. This is to say that equilibrium thermodynamics itself does not provide a broad enough basis that will incorporate all types of kinetic process.

This book is formalistic rather than applicative in character. My feeling is that the presentation of a self-consistent and clear-cut thermodynamic formulation will automatically lead to its application. I hope that the interdisciplinary character of the book will allow the reader to draw upon the many interesting analogies that exist among the seemingly diverse branches of macroscopic physics, chemistry and theoretical biology.

I would like to express my deep gratitude to Gabriel Stein of the Hebrew University, Jerusalem, whose encouragement and advice had a great deal to do with making this book a reality. A very special role has been played by my wife Fanny, to whom I am greatly indebted.

Lucrino, Italy BERNARD H. LAVENDA

Introduction

Formulations of non-equilibrium thermodynamics

In the first part of this book (chapters 1–4), there is given a brief exposition of classical equilibrium and linear thermodynamics and a critical review of two recent formulations of non-equilibrium thermodynamics.

Chapter 1 briefly presents the classic formulations of equilibrium thermodynamics. There is a twofold objective: (1) to accentuate the inherent differences between the axiomatic formulation of Carathéodory and the phenomenological approach of Gibbs, and (2) to evaluate the relative merits of the two equilibrium formulations for the construction of a non-equilibrium theory. The conclusion is reached that although Carathéodory's theory is mathematically more rigorous, it lacks the elements which would make it readily adaptable as a basis for the development of a theory of non-equilibrium thermodynamics.

Chapter 2 gives a chronological account of the developments in linear thermodynamics. Onsager's derivation of a class of reciprocal relations in which the flux is the time derivative of an extensive thermodynamic variable, forms the corner-stone of linear thermodynamics. The advantage of placing Onsager's derivation in its historical original form is that it affords a better grasp of the reciprocal relations and their specificity. An objection is raised concerning the Onsager – Casimir demonstration, in that their interpretation of the principle of microscopic reversibility at equilibrium is apparently incongruous with the non-conservative nature of the phenomenological regression laws.

Chapter 3 presents a résumé of one school of thought which uses a formal statement of the second law, the so-called Clausius – Duhem inequality, as a restriction on the types of thermodynamic process that can occur in elastic materials. Why this restriction? Since the second law does not, in general, constitute a criterion of stability, this restriction would necessarily exclude various forms of nonlinear thermodynamic processes which are stable kinetically. In the last section of this chapter there is an example of where the restriction on the form of the constitutive relations leads to a contradiction; the results invalidate the Clausius – Duhem inequality.

Chapter 4 is a review of the work of still another school of thought, which uses the sign criteria of the second variation of the entropy and its time-rate-of-change as criteria of stability in the small. The inability to make a direct connection with the second law, on account of the fact that the first variation of the entropy does not vanish in a non-equilibrium stationary state, makes it necessary to turn to a weaker justification of the proposed stability criteria based on an analogy with a Liapounov function. Notwithstanding the fact that Liapounov's second method is addressed to stability in the large, the analogy is found to be spurious. One of the two criteria of Liapounov's second method is satisfied automatically by supposing the system to be in a state of local equilibrium. The criteria of local equilibrium have nothing whatsoever to do with the asymptotic stability properties of kinetic processes. Furthermore, it is shown that the sign criterion of the time-rate-of-change of the second variation of the entropy does not coincide with the necessary and sufficient conditions of stability that are obtained from Liapounov's first method.

Quasi-thermodynamic approach to nonlinear thermodynamics

In the second part of the book (chapters 5–9), there is undertaken a detailed exposition of an approach to nonlinear thermodynamics which is based on a confluence of thermodynamic and kinetic concepts regarding evolution and stability. For the major part, the analyses are limited to thermodynamic systems that are found in the immediate neighbourhood of a non-equilibrium stationary state. Only in the last sections of chapters 7 and 9 is the approach extended to include the phenomena of nonlinear periodic processes in space and time that may occur at a finite distance from an unstable non-equilibrium stationary state.

The fundamental idea is that the principles of thermodynamics are compatible with, and can be sharpened by, the precise criteria of kinetic stability analysis. In carrying out the implications of the fundamental idea, it was found necessary to correlate thermodynamic variables, needed to specify the thermodynamic state, with mechanical variables that satisfy the same types of differential equation. This approach is to be regarded as 'quasi-thermodynamic' in character and this is what distinguishes it from the thermodynamic approaches that have been discussed in the first part of the book.

In chapter 5, the development of nonlinear thermodynamics is begun, having already appreciated the fact that linear thermodynamic processes evolve to a state of least dissipation of energy or equivalently to a state of minimum entropy production. This implies that the evolution of linear thermodynamic processes can be accounted for by a single thermodynamic potential. The relevant thermodynamic principle states that the entropy production is equal to the energy dissipated.

In nonlinear thermodynamics we are dealing with processes that occur in open systems in which the external forces prevent the system from relaxing to equilibrium. We can no longer expect that the linear thermodynamic principle will be valid or that the evolution of such processes can be accounted for in terms of the properties of a single thermodynamic potential. It is found necessary to derive an extension of the thermodynamic principle which will govern the evolution of nonlinear thermodynamic processes.

It is now found that the dissipation function is no longer synonymous to the entropy production but rather that their difference is a measure of the absorbed power. This is to say, the absorbed power appears as the time-rate-of-change of the entropy less that which is dissipated. The thermodynamic principle of the balance of power forms the basis for the discussion of non-equilibrium variational principles in chapter 6 and the quasi-thermodynamic stability analysis of chapter 7.

In chapter 6, there is a fairly complete treatment of the variational principles of linear and nonlinear thermodynamics. How non-equilibrium thermodynamic variational principles are constructed is shown by applying constraints on the principle of least dissipation of energy. In linear thermodynamics, the principle of minimum entropy production is obtained as a corollary of the aforementioned principle and does not constitute a minimum principle in itself.

Throughout, this chapter deals with a 'conditioned' rather than a free minimum of the dissipation function. In other words, we are not merely interested in determining the values of the velocities for which the dissipation function becomes a minimum but we are asked, at the same time, to consider only those values of the velocities for which the nonlinear thermodynamic principle of the balance of power is simultaneously fulfilled.

An appreciation is gained of the relation between thermodynamic variational principles and those of classical mechanics. The thermodynamic variational principle is analogous to the classical Lagrangian function whose stationary value yields the phenomenological equations corresponding to the classical equations of motion. Rather than obtaining the principle of the conservation of energy through integration, we multiply the phenomenological equations by their conjugate velocities and sum them to obtain the thermodynamic principle of the balance of power.

In chapter 7, there is raised the question of whether thermodynamic stability criteria can supplement the necessary and sufficient conditions for asymptotic stability which are obtained from Liapounov's first method. The precise but particular results of Liapounov's first method are employed and general thermodynamic criteria of stability under certain circumstances are obtained.

The advantages of the quasi-thermodynamic analysis are: (1) it may provide general criteria for asymptotic stability that are not manifested by the integration of the variational equations, and (2) the stability analysis, under certain conditions, may be reduced to the application of definite criteria that have a thermodynamic significance. It is found that in all cases where the thermodynamic force is conservative, the necessary and sufficient conditions for asymptotic stability are that the dissipation function and the second variation of the entropy must be positive and negative definite, respectively. The vanishing of either quadratic form means that the stability is critical.

In chapter 7, Liapounov's first method is used to extract definite thermodynamic stability criteria from the complex power equation. Attention is then turned to the symmetries of the phenomenological coefficient matrices. These symmetries determine whether or not the exactness conditions are satisfied which guarantee the existence of scalar thermodynamic potentials. In turn, by means of a mechanical analogy, the symmetries are related to conservative or non-conservative forces. The quasi-thermodynamic analysis provides a classification of the various forms of system motions in terms of their influence on system stability and prepares the way for the field analyses of the following chapter.

In chapter 8, methods to analyse nonlinear thermodynamic processes which are non-conservative are developed, recalling from the preceding chapter that the only case in which definite thermodynamic criteria of stability cannot be obtained is when a non-conservative thermodynamic force exists. In this chapter the situation is remedied.

The field analysis is particularly well adapted to two types of critical nonlinear thermodynamic phenomenon: (1) multistationary state transitions which are signalled by the vanishing of the second variation of a scalar potential called the velocity potential, and (2) rotational processes that are caused by non-conservative thermodynamic forces.

Methods are developed that use both velocity and force fields. When the system motion can be effectively reduced to a half-degree of freedom, the velocity potential method determines bifurcation points that are analogous to equilibrium critical points in second order phase transitions. Rotational processes require at least a single degree of freedom and the velocity field analysis can only describe the kinematics and not the dynamics of the motion.

The dynamics of rotational processes are obtained from a thermodynamic force field analysis which parallels the classical development of macroscopic field theories. The only difference is that real space is replaced by configuration space. Non-conservative forces are now shown to be derivable from thermodynamic vector potentials. Thermodynamic field principles and equations that govern the field are obtained from the constrained principle of least dissipation of energy. The new factor in the thermodynamic field principles is the presence of an energy flux density which implicates an external energy source in the maintenance of nonlinear thermodynamic rotational processes.

In chapter 9, there is formulated an internal state variable representation of thermodynamic processes belonging to the continua. After the derivation of a general thermodynamic principle of the continua, we limit ourselves to the analysis of isothermal processes under constant strain. These processes are governed by the following thermodynamic principle: the difference between the absorbed power and the energy flux across the surface of the system appears as the time-rate-of-change of the entropy less the energy which is dissipated. The thermodynamic principle allows non-equilibrium thermodynamic variational principles of the continua to be formulated and thus both the variational equations and the relevant thermodynamic stability criteria are obtained.

Considered next are thermodynamic evolutionary criteria that are derived by means of the maximum principles of partial differential equations. It is appreciated that the internal configuration of continuous systems can be determined uniquely and solely from an analysis of the flows across their boundaries. A simple example is a system without any heat sources: the temperature at any point in the system cannot be greater than the temperature at the surface. Any heat flow resulting from a difference in temperature will be directed into the system. It is shown that a positive directional derivative of the entropy flux at the surface of the system provides the condition for the absence of a positive maximum or a negative minimum in the spatial distribution of the variation of an internal state variable. These spatial distributions cause a decrease in the system entropy which must be compensated by an entropy influx from the surroundings.

Chapter 9 is concluded by an extension of the quasi-thermodynamic approach to the analysis of truly nonlinear processes. In much the same spirit that thermodynamic and kinetic concepts were interrelated in chapter 7, an averaged thermodynamic variational principle that employs the averaging techniques of nonlinear mechanics is now formulated. The method is directed to the analysis of the asymptotic evolution of nonlinear dissipative wavetrains and it is found, under certain conditions, that they give rise to a limit wave phenomenon. This brings us to the present-day frontiers of nonlinear thermodynamics.

1 Equilibrium Thermodynamics

The purpose of this chapter is to compare the two different formulations of equilibrium thermodynamics only in so far as they bear on the theories of linear and nonlinear thermodynamics. The two formulations are generally associated with the names of Carathéodory (1909, 1925) and Gibbs (1902). Carathéodory's theory is axiomatic in character, whereas Gibbs's theory is phenomenological. Although it has been suggested that Carathéodory's theory may eventually provide a better mathematical basis for a theory of non-equilibrium thermodynamics (Eckart, 1940), most of the recent theories of non-equilibrium thermodynamics possess a predominantly Gibbsian character.

Equilibrium thermodynamics is concerned with (1) the definition and proof of existence of the entropy and absolute temperature, and (2) the methodology by which equilibrium properties of thermodynamic systems can be characterised. Carathéodory's theory is concerned primarily with the axiomatisation of (1), while Gibbsian thermodynamics is directed to (2). Although the two formulations of equilibrium thermodynamics are consistent with one another, they differ in their aims and points of view. There is no *a priori* reason why we should select one over the other as a basis for non-equilibrium thermodynamics.

The classical theory of equilibrium thermodynamics is therefore a blend of two essentially different approaches. Carathéodory's theory is a mathematical formulation of the empirical results of Clausius (1850) and Kelvin (Thomson, 1848). It sought to replace thermodynamic arguments based on Carnot cycles by a differential geometry dealing with Pfaffian differential forms. In other words, Carathéodory's theory allows us to obtain all the mathematical consequences of the second law of thermodynamics without recourse to any particular physical model (Chandrasekhar, 1939). One of the major achievements of Carathéodory's theory is the definition and proof of existence of the entropy solely in terms of mechanical variables, such as pressure and volume. Carathéodory's axiomatic approach was developed and expounded upon by Born (1921). Although the theory has been frequently discussed in the literature (Born, 1949; Buchdahl, 1949, 1954, 1955; Chandrasekhar, 1939; Eisenschitz, 1955; Landé, 1926; Margenau and Murphy, 1943) there have been no substantial advances made since 1921, with the important exception of Landsberg's (1956) geometrisation of Carathéodory's axiomatic approach.

Carathéodory's theory is not concerned with the characterisation of the equilibrium properties of thermodynamic systems. In Gibbsian thermodynamics, on the other hand, attention is focused on the system. The primitive concepts of entropy and absolute temperature are assumed not to be needing definition or existence of proof. It uses these concepts to obtain a more detailed description of equilibrium thermodynamic systems in terms of their chemical and phase structures. One of the major advantages of Gibbsian thermodynamics, over Carathéodory's theory, is that it provides criteria for its validity. One assumes, in Carathéodory's theory, that all transformations are non-singular, whereas in Gibbs's theory the singularities of certain transformations are shown to coincide with the limits of system stability (e.g. critical points). However, one of the major disadvantages of Gibbsian thermodynamics is that an axiomatic approach, comparable to Carathéodory's theory, is still lacking (cf. Tisza, 1966).

The two formulations of equilibrium thermodynamics can be discussed in terms of the geometrical structure of their relevant spaces. A major tool in equilibrium thermodynamics is the geometrical method based on the 'thermodynamic' space. In Carathéodory's theory, the space is spanned by all the independent variables that are needed to describe the thermodynamic system. The only criterion for their presence in the thermodynamic space is that they must be physically measurable. Hence, in Carathéodory's formulation no specific choice of the space is made; in particular, no geometrical distinction is made between extensive and intensive thermodynamic variables. This was first pointed out by Ehrenfest (1911), who concluded that the geometrical distinction between extensive and intensive variables requires additional axioms than those contained in Carathéodory's theory. The geometrisation of Carathéodory's theory was accomplished by Landsberg (1956, 1961).

In contrast, Gibbsian thermodynamics requires a particular space—the Gibbs space—which is spanned by all the independent extensive variables. Although in the original presentation (Gibbs, 1902), the construction of the Gibbs space was *ad hoc*, it did lead to the representation of thermodynamic potentials as quadratic forms which form the primitive surfaces in the Gibbs space. The achievements of Gibbs's geometrisation of his thermodynamic theory is really quite remarkable when it is realised that the basic geometrical elements of metric and orthogonality are wanting.

The mathematical foundations were discovered, at a later date, by Blaschke (1923) who showed that the geometry of the Gibbs space is an affine differential geometry. Although a metric cannot be defined, it is possible to define a parallel projection which replaces the orthogonal projection in the ordinary Riemann theory of curvature. Furthermore, by representing the entropy as a quadratic form, it is possible to obtain something similar to a metric. That is, the Gibbs space is one in which 'volumes' but not 'lengths' are measurable. The volume is represented by the determinant of the matrix that is associated with the quadratic form. Therefore, in contrast to Euclidean geometry, parallelism replaces orthogonality, and volumes but not lengths are measurable in the Gibbs space. In addition, different spaces can be generated that are spanned by both extensive and intensive variables, through the Legendre transforms which replace extensive variables by their conjugate intensive variables in the fundamental expression for the thermodynamic potential.

The attributes of the axiomatic and phenomenological formulations of equilibrium thermodynamics are now considered in greater detail. Since the laws of thermodynamics are formulated with the greatest amount of logical simplicity in Carathéodory's theory, his formulation will be considered first. There will then follow a discussion of Gibbsian thermodynamics which takes many of the concepts, introduced by Carathéodory, and makes practical use out of them.

1.1 Carathéodory's theory

1.1.1 Definitions and conditions of equilibrium

Thermodynamic systems are classified according to whether they are 'isolated', 'closed' or 'open'. In an isolated system there is no communication with the outside world; that is, there is neither energy nor matter transfer between the system and its environment. In a closed system there is energy but no matter exchange with the outside world. Finally, in an open system there is both an exchange of energy and matter between the thermodynamic system and its environment.

Since equilibrium thermodynamics deals with isolated systems, how is it possible to establish the conditions of equilibrium? In other words, if the isolated system is not in equilibrium then there is the problem of defining a thermodynamic potential for the non-equilibrium state. Alternatively, if the system is in equilibrium it is no longer possible to determine the conditions of equilibrium. To resolve this paradoxical problem, Carathéodory introduced the ingenious device of 'composite systems' which was later carried over into Gibbsian thermodynamics (cf. section 1.2). That is, in order to obtain the conditions of equilibrium, in a theory that does not define non-equilibrium states, it is necessary to introduce partitions that divide the isolated system into a number of subsystems. The partition or 'wall' permits the passage of a certain form of energy which thereby forces a new relation upon the parameters of the subsystems that it separates (Landsberg, 1956).

Carathéodory's theory considers only two types of partition that are involved with the transfer of heat: (1) an adiabatic partition which is restrictive to the passage of thermal energy, and (2) a diathermal partition which allows the passage of thermal energy. Diathermal partitions are used to establish the conditions of thermal equilibrium. But as Landsberg (1956) points out, the definition of a diathermal partition precedes the introduction of the notions of heat and thermal equilibrium in the axiomatic formulation. This is only one of a number of criticisms that can be lodged against Carathéodory's formulation (cf. Landsberg, 1956).

In Gibbsian thermodynamics, the classes of partitions are increased so that other criteria of equilibrium can be obtained in addition to those of thermal equilibrium. Both adiabatic and diathermal partitions are impermeable to the transfer of matter. The two other types of partition, that are considered in Gibbsian thermodynamics, are: (3) a semipermeable partition that restricts the passage of certain constituents, and (4) a permeable partition which is non-restrictive to the passage of matter. Both these partitions permit the exchange of thermal energy between the subsystems. It can be appreciated that partitions (3) and (4) establish the conditions of chemical equilibrium only after the system has attained thermal equilibrium. This is entirely reasonable since energy, but not matter, can be exchanged, the converse of which is certainly not true.

In order to obtain the conditions of thermal equilibrium, an isolated system is divided into a number of subsystems by means of diathermal partitions. For simplicity, consider the case of a single diathermal partition which separates two subsystems. Each of the subsystems is assumed to be fully characterised in terms of their pressures, p and p', and their volumes, V and V'. In order for there to exist an equilibrium between the two subsystems, the four parameters must enter into a relation of the form

$$F(p, V, p', V') = 0 \qquad (1.1.1)$$

that depends only on the properties of the two subsystems. In Carathéodory's theory, (1.1.1) is the condition of thermal equilibrium. The condition of thermal equilibrium may be extended to any number of subsystems by the transitive property: if two subsystems are in equilibrium with another subsystem, then they will be in thermal equilibrium with each other when they are brought into thermal contact. This means that the condition of thermal equilibrium has the form

$$T(p, V) - T'(p', V') = 0 \qquad (1.1.2)$$

where T and T' are the empirical temperatures of the two subsystems. Thermal equilibrium then implies the equivalence of the empirical temperatures, namely $T = T'$.

Another set of definitions concerns the types of thermodynamic process that can be performed. A 'quasi-static' process is one that is carried out infinitely slowly so that the system can be considered as passing through a continuous series of equilibrium states. Here there exists an ambiguity in the definitions of a quasi-static process and a 'reversible' process (Landsberg, 1956). The two types of process are generally regarded as being synonymous. Any other process is called 'non-statical' which has the connotation of being 'irreversible'.

1.1.2 The first law

The first law of thermodynamics is nothing more than the principle of conservation of energy that applies to systems where heat is produced or absorbed (Planck, 1954). In Carathéodory's theory, the internal energy is defined solely in mechanical terms in contrast to the concept of heat which is a derived one, having no independent significance apart from the first law (Chandrasekhar, 1939).

In an adiabatic process which brings the system from an initial state (1) to a final state (2), the work done on the system is equal to the increase in its internal energy, namely

$$E_2 - E_1 = \int_1^2 dW \qquad (dQ = 0) \qquad (1.1.3)$$

The internal energy E is a function of state; it depends only on the initial and final states. If the transition $1 \to 2$ were to occur along a non-adiabatic path ($dQ \neq 0$), the work W would depend upon the path. For a non-adiabatic path, the quantity of heat supplied to the system is

$$\int_1^2 dQ = E_2 - E_1 - \int_1^2 dW \tag{1.1.4}$$

It is a remarkable fact that, whereas the work done on the system and the heat supplied to the system are not point functions, their sum, the internal energy, is a point function. This means that dW and dQ are not 'exact' or 'perfect' differentials of scalar functions. Only their sum

$$dE = dW + dQ \tag{1.1.5}$$

is a perfect differential. In physical terms, we can attribute a certain internal energy with a given thermodynamic state but we cannot speak about the quantity of heat that the system possesses in that state. Moreover, if two subsystems are separated from one another by an adiabatic partition, then by definition

$$E = E_1 + E_2 \tag{1.1.6}$$

Expressed in words, the internal energy of the system is equal to the sum of its components. This will, in general, not be true when two subsystems are brought into contact. There will then be an additional energy due to the contact. The contact energy is proportional to the surface area in common between the subsystems, so that if the volume to surface ratio is large, as might be expected in all thermodynamic systems, then the additional energy is negligible and we suppose the additive law (1.1.6) always to be valid.

1.1.3 The second law

The first law of thermodynamics does not provide for a unique determination of physical processes. For example, the conservation of energy cannot tell us in which direction the heat is flowing between hot and cold bodies. In other words, the first law does not determine the direction in which a process takes place. From the viewpoint of the first law, initial and final states are entirely equivalent (Planck, 1954).

The second law of thermodynamics can be considered a law of 'impossibility'. It tells us which processes are physically unrealisable. The empirical basis of the second law is found in the Clausius–Kelvin principle. One phrasing of the second law is: without 'compensation', it is impossible to transfer heat from a cold to a hot body (Chandrasekhar, 1939). The mathematical formulation of the second law is due to Carathéodory, who showed that the absolute temperature is an integrating denominator for dQ. The second law can then be expressed as the inability to reach states arbitrarily near to a given state by means of an adiabatic process. In order to appreciate Carathéodory's theorem it is necessary to consider the mathematical properties of Pfaffian differential equations (Chandrasekhar, 1939; Goursat, 1922; Margenau and Murphy, 1943; Schouten and van der Kulk, 1949).

Consider a Pfaffian differential expression in two variables

$$dQ = X(x, y)dx + Y(x, y)dy \tag{1.1.7}$$

If

$$\int_1^2 dQ \neq Q(x_2, y_2) - Q(x_1, y_1) \tag{1.1.8}$$

then dQ is not the true differential of a certain function. This means that dQ cannot be integrated; it depends on the path from $1 \rightarrow 2$. For if dQ were a perfect differential of some scalar function f, with $dQ = df$, we would have

$$df = \partial_x f dx + \partial_y f dy \tag{1.1.9}$$

Comparing (1.1.7) and (1.1.9) we find

$$X(x, y) = \partial_x f \qquad \text{and} \qquad Y(x, y) = \partial_y f \tag{1.1.10}$$

or

$$\partial_y X = \partial_x \partial_y f = \partial_x Y \tag{1.1.11}$$

which is the exactness or integrability condition for the scalar function f. Condition (1.1.11) is, in general, not satisfied.

The Pfaffian differential equation, corresponding to (1.1.7), is

$$dQ = X dx + Y dy = 0 \tag{1.1.12}$$

Equation (1.1.12) defines a direction in the tangent plane to the one parametric family of curves which are the solutions to (1.1.12) at any given point. This is to say that we can equally as well write (1.1.12) as

$$d_x y = -X / Y \tag{1.1.13}$$

which is tangent to the curve

$$f(x, y) = C = \text{const} \tag{1.1.14}$$

at a given point. From (1.1.14) it follows that $d_x f = 0$ which is expressed explicitly as

$$\partial_x f + \partial_y f d_x y = 0 \tag{1.1.15}$$

Inserting (1.1.13) into (1.1.15), we obtain

$$\partial_x f - \partial_y f X / Y = 0 \tag{1.1.16}$$

or

$$Y \partial_x f = X \partial_y f = XY/\Theta, \text{ say} \tag{1.1.17}$$

where Θ is a function of x and y. (If $dQ = df$ then $\Theta = 1$.) We can now write (1.1.17) as

$$X = \Theta \, \partial_x f \qquad \text{and} \qquad Y = \Theta \, \partial_y f \tag{1.1.18}$$

Then the substitution of (1.1.18) into (1.1.7) leads to

$$dQ = \Theta(\partial_x f dx + \partial_y f dy) = \Theta \, df \tag{1.1.19}$$

In other words, if we divide the Pfaffian expression (1.1.7) by the integrating denominator Θ we obtain an exact differential. This is not surprising since a Pfaffian expression in two variables will always admit an integrating denominator. Moreover, if a Pfaffian differential expression admits one integrating denominator, it will admit an infinity of them. We can replace f by any function $S = S[f(x, y)]$ for which

$$S[f(x, y)] = C = \text{const} \tag{1.1.20}$$

are solutions to the Pfaffian differential equation. Then

$$dS = d_f S df = d_f S dQ/\Theta = 1/\theta \, dQ \tag{1.1.21}$$

with

$$\theta(x, y) = \Theta \, d_S f \tag{1.1.22}$$

For Pfaffian differential expressions in more than two variables, integrating denominators can be found only in very particular cases. Equilibrium thermodynamics is such a particular case. Therefore, Pfaffian differential expressions can be categorised according to whether they admit or do not admit integrating denominators. Physically what does this mean? If the two variable case is considered, then through every point in the x, y plane there passes one and only one curve of the family (1.1.14). Thus, the existence of an integrating denominator means that we cannot reach all neighbouring points to a given point by means of curves that satisfy the Pfaffian differential equation (1.1.12). If dQ is associated with the heat transfer then (1.1.21) is significant of the fact that there exists states which are inaccessible to a state along (1.1.14) by means of an adiabatic process.

The opposite question can now be asked: if there exist states that are inaccessible to those which lie along curves which are solutions to the Pfaffian differential equation then does an integrating denominator exist? Carathéodory answered this question in the affirmative and it is known as Carathéodory's theorem. The proof is essentially as follows: consider the case of a Pfaffian differential expression in three variables. There are three possibilities for states in the neighbourhood of a given state γ_0: (1) they fill a certain volume enclosing γ_0, (2) they lie on a surface element containing γ_0, or (3) they are on a curve passing through γ_0. The first possibility is rejected since it contradicts the assumption that not all states, however near to a given state, are accessible to it. The third possibility is also rejected on account of the fact that

$$dQ = X dx + Y dy + Z dz = 0 \tag{1.1.23}$$

defines a surface element that contains states accessible to γ_0 by means of an adiabatic process. Therefore, the points accessible to γ_0 must comprise a surface element $d\Gamma_0$.

Now consider the states on the boundary of $d\Gamma_0$, namely γ'. We can construct a new surface element $d\Gamma'$ that contains all those states which are accessible to γ'. Consideration is now given to the relative orientations of these two surfaces. If $d\Gamma'$ would intersect $d\Gamma_0$ then it would be possible to reach any neighbouring state, say γ'', by an adiabatic process by first proceeding from γ_0 to γ' along $d\Gamma_0$ and then from γ' to γ'' along $d\Gamma'$. This contradicts our hypothesis, so the remaining possibility is that $d\Gamma_0$ and $d\Gamma'$ form a continuous set of surface elements. If γ'' is inaccessible to γ_0 then it must lie on another surface element which does not intersect or touch the surface element $d\Gamma_0$. Hence, the space is filled with an entire family of non-intersecting surfaces.

Carathéodory's theorem can now be stated as: if a Pfaffian differential expression

$$dQ = X dx + Y dy + Z dz \tag{1.1.24}$$

possesses the property that there are states that cannot be connected to a given state along curves that satisfy (1.1.23), then (1.1.24) admits an integrating denominator, namely

$$dQ = \theta(x, y, z)\,dS \tag{1.1.25}$$

Herein lies the substance of the second law of thermodynamics. First consider the case of an adiabatic process that is quasi-static. Applying Carathéodory's theorem to (1.1.12) is superfluous since it is known that in the two variable case there will always exist an integrating denominator. The full strength of Carathéodory's theorem is felt when we consider a composite system.

Carathéodory's theorem asserts that

$$dQ = dQ_1 + dQ_2 \tag{1.1.26}$$

can always be expressed in the form

$$dQ = \Theta(x_1, x_2, T)\,df(x_1, x_2, T) \tag{1.1.27}$$

where x_1 and x_2 are mechanical variables that characterise each of the subsystems. For each of the subsystems we have

$$dQ_1 = \Theta_1(x_1, T_1)\,df_1(x_1, T_1) \tag{1.1.28}$$

$$dQ_2 = \Theta_2(x_2, T_2)\,df_2(x_2, T_2) \tag{1.1.29}$$

When the subsystems are brought into thermal contact, $T_1 = T_2 = T$. On account of (1.1.26) and (1.1.27) we can write

$$\Theta df = \Theta_1 df_1 + \Theta_2 df_2, \tag{1.1.30}$$

We are at liberty to choose f_1, f_2 and T as the independent variables so that from (1.1.30) we obtain

$$\begin{aligned} \partial_{f_1} f &= \Theta_1(f_1, T)/\Theta(f_1, f_2, T) \\ \partial_{f_2} f &= \Theta_2(f_2, T)/\Theta(f_1, f_2, T) \\ \partial_T f &= 0 \end{aligned} \tag{1.1.31}$$

From the third equation in (1.1.31) we conclude that

$$f = f(f_1, f_2) \tag{1.1.32}$$

and, consequently, the dependent variables in equation (1.1.30) must also be independent of T, namely

$$\partial_T(\Theta_1/\Theta) = \partial_T(\Theta_2/\Theta) = 0 \tag{1.1.33}$$

Performing the differentiation explicitly, we find

$$\partial_T \log \Theta_1 = \partial_T \log \Theta_2 = \partial_T \log \Theta = \alpha(T), \text{ say} \tag{1.1.34}$$

where $\alpha(T)$ is a universal function of the empirical temperature. It can only be a function of T because it is the only thing in common to both subsystems.

Integrating (1.1.34) we obtain

$$\log \Theta_i = \int \alpha(T)\,dT + \log F_i(f_i), \qquad i = 1, 2 \tag{1.1.35}$$

$$\log \Theta = \int \alpha(T)\,dT + \log F(f_1, f_2) \tag{1.1.36}$$

where the constants of integration F_i and F are necessarily independent of the empirical temperature. Equations (1.1.35) and (1.1.36) can be written in the form

$$\Theta_i = F_i(f_i)\theta/K, \qquad i = 1, 2 \tag{1.1.37}$$

$$\Theta = F(f_1, f_2)\theta/K \tag{1.1.38}$$

where the absolute temperature θ is defined as

$$\theta = K \exp \int \alpha(T) \mathrm{d}T \tag{1.1.39}$$

K being an arbitrary constant. Introducing (1.1.38) into (1.1.27) leads to

$$\mathrm{d}Q = \Theta \mathrm{d}f = (\theta/K)F \mathrm{d}f \tag{1.1.40}$$

We now define the 'entropy' of the subsystems as

$$S_i = (1/K)\int F_i(f_i)\mathrm{d}f_i + \text{const.}, \qquad i = 1, 2 \tag{1.1.41}$$

The entropy of each subsystem is a function only of the variables characterising it; we observe that it is independent of the temperature. Moreover, it is definable to within an arbitrary constant. For each of the separate subsystems we have

$$\mathrm{d}Q_i = \Theta_i \mathrm{d}f_i = (\theta/K)F_i(f_i)\mathrm{d}f_i = \theta \mathrm{d}S_i, \qquad i = 1, 2 \tag{1.1.42}$$

while for the composite system

$$\mathrm{d}Q = (\theta/K)[F_1(f_1)\mathrm{d}f_1 + F_2(f_2)\mathrm{d}f_2] \tag{1.1.43}$$

or, from (1.1.40),

$$F(f_1, f_2)\mathrm{d}f = F_1(f_1)\mathrm{d}f_1 + F_2(f_2)\mathrm{d}f_2 \tag{1.1.44}$$

Hence

$$F(f_1, f_2)\partial_{f_1} f = F_1(f_1) \tag{1.1.45}$$

$$F(f_1, f_2)\partial_{f_2} f = F_2(f_2) \tag{1.1.46}$$

Differentiating (1.1.45) and (1.1.46) with respect to f_2 and f_1, respectively, and subtracting, we find that the Jacobian

$$\frac{\partial(f_1, F)}{\partial(f_1, f_2)} = 0 \tag{1.1.47}$$

This means that F is functionally related to f and only depends on f_1 and f_2 through f, i.e.

$$F(f_1, f_2) = F(f) \tag{1.1.48}$$

Thus we can define the total entropy of the composite system as

$$S = (1/K)\int F(f)\mathrm{d}f + \text{const} \tag{1.1.49}$$

and by a suitable choice of the arbitrary constant we have that the entropy is the sum of the entropies of the subsystems.

Then for a quasi-static process,

$$(1/\theta)\mathrm{d}Q = \mathrm{d}S \tag{1.1.50}$$

which is the well-known form of the second law. Expression (1.1.50) can now be compared with (1.1.27). Whereas Θ and f depend on all the physical variables in (1.1.27), there is a separation of the functional dependencies in (1.1.50): θ is a function only of the empirical temperature which is uniform throughout the isolated system, while S is a function of the variables f_1 and f_2 which are constant for adiabatic changes.

Let us now generalise our results to non-statical processes. Hitherto, we have considered that the thermodynamic system is characterised by the variables x_1, x_2 and T, cf. expression (1.1.27). Since Carathéodory's theory is insensitive to the explicit choice of independent variables, we can equally as well choose x_1, x_2, and S. Consider the transition between the states (x_1, x_2, S) and (x_1', x_2', S') to occur in two stages. The first stage consists of the change $x_1, x_2 \to x_1', x_2'$, which is carried out by a quasi-static, adiabatic process with $S = \text{const}$. In the second stage, we hold x_1' and x_2' constant and perform an adiabatic process ($dQ = 0$), which is non-statical ($dQ \neq \theta dS$), that changes S into S'. If the entropy increases in certain cases while decreasing in others, it could always be arranged that any neighbouring state of a given initial state could be reached by an adiabatic process. This contradicts Carathéodory's principle in its more general form, which allows for non-statical processes.

As a consequence, the entropy must be a monotonic function; it must always increase or always decrease. To find out which is in fact the case, an appeal is made to an ideal gas experiment. It is then concluded that the entropy can never decrease during an adiabatic, non-static change. This fact can be expressed in a different way by considering a cyclical transition between two states. The forward transition $1 \to 2$ occurs adiabatically (but not necessarily statically), whereas the reverse transition $2 \to 1$ occurs quasi-statically. We integrate (1.1.50) over the complete cycle and write

$$\oint dQ/\theta = \int_1^2 dQ/\theta + \int_2^1 dQ/\theta \tag{1.1.51}$$

The first member in (1.1.51) vanishes since we have carried out the process adiabatically. The second member is just the entropy difference $(S_1 - S_2)$ and since the entropy must increase (or at most remain constant) we have

$$\oint dQ/\theta \leq 0 \tag{1.1.52}$$

In this form, Clausius first enunciated the second law.

If we combine the first and second laws of thermodynamics, (1.1.4) and (1.1.52), respectively, some interesting results are obtained. Since the transition $2 \to 1$ occurs quasi-statically, we have from (1.1.51) and (1.1.52) that

$$\int_1^2 dQ/\theta \leq S_2 - S_1 \tag{1.1.53}$$

Suppose that the process is isothermal, i.e.

$$\int_1^2 dQ \leq \theta(S_2 - S_1) \tag{1.1.54}$$

Introducing (1.1.4) into (1.1.54) results in

$$F_2 - F_1 \leq \int_1^2 dW \qquad (\theta = \text{const}) \tag{1.1.55}$$

where F is the Helmholtz free energy

$$F \equiv E - \theta S \tag{1.1.56}$$

From (1.1.55) we conclude that in an isothermal process in which no work is done on the system, the Helmholtz free energy can never increase.

As pleasing as Carathéodory's theory is from a mathematical viewpoint, it does have shortcomings that do not make it readily applicable to the analysis of physical processes. A major shortcoming is that it does not provide the criteria for its validity. This is not true of the more phenomenologically oriented theory of Gibbs.

1.2 Gibbsian thermodynamics

The starting point of Gibbsian thermodynamics is the elimination of dQ between equations (1.1.5) and (1.1.50)

$$dS = (1/\theta)dE + (p/\theta)dV \tag{1.2.1}$$

which is obviously valid for a quasi-static or reversible process. Equation (1.2.1) applies to the special case in which there is only compressional work done, namely

$$dW = -p\,dV \tag{1.2.2}$$

It has already been mentioned that in Gibbsian thermodynamics one assumes that the entropy and absolute temperature are neither in need of definition nor proof of existence. So it would appear that Gibbsian thermodynamics begins where Carathéodory's theory leaves off.

The independent variables in equation (1.2.1) are the internal energy and the volume. All that is needed experimentally is a thermometer and a pressure gauge to derive the relations

$$1/\theta = g_1(E,\ V) \tag{1.2.3}$$

$$p/\theta = g_2(E,\ V) \tag{1.2.4}$$

Expressions (1.2.3) and (1.2.4) are characteristic of the specific thermodynamic system under investigation. They are commonly referred to as 'equations of state'. If the transformation is non-singular (such as at critical points), the roles of the dependent and independent variables can be inverted. We then obtain the 'thermal'

$$V = V(p,\ \theta) \tag{1.2.5}$$

and 'caloric'

$$E = E(p,\ \theta) \tag{1.2.6}$$

equations of state.

The two equations of state (1.2.3) and (1.2.4) are not independent of one another. They satisfy the relation, cf. condition (1.1.11),

$$(\partial_V 1/\theta)_E = (\partial_E p/\theta)_V \tag{1.2.7}$$

which is recognised as the exactness or integrability condition for the entropy. Condition (1.2.7) implies that equation (1.2.1) can be integrated to give

$$S = S(E,\ V) \tag{1.2.8}$$

which is the so-called 'fundamental' equation (Tisza, 1966). Once equation (1.2.8) has been obtained, the equations of state follow from the differentiation of (1.2.8), namely

$$(\partial_E S)_V = 1/\theta \tag{1.2.9}$$

$$(\partial_V S)_E = p/\theta \tag{1.2.10}$$

Consequently, the knowledge of both equations of state (1.2.3) and (1.2.4) is tantamount to all that is known about the thermodynamic system. All this information is contained in equation (1.2.8), hence the name 'fundamental' equation.

In Gibbsian thermodynamics, attention is focused on the system. The distinction between extensive and intensive variables is now made clear: the thermodynamic system is characterised by extensive variables, whereas the coupling between the system and environment is described by the intensive variables or 'intensities'. One of Gibbs's major achievements was the generalisation of the fundamental equation to include other types of extensive variable that are associated with the performance of work. However, in Gibbsian thermodynamics not all types of work can be accounted for by the fundamental equation. Since the thermodynamic system is isolated, we have the criterion that the extensive variables must be 'additive invariants'. This is to say that we consider a composite system where, for example, the internal energy in subsystem k is $E^{(k)}$. The energies of the different subsystems are additive and, on account of the restrictive enclosures, the sum is required to be a constant, namely

$$\sum_k E^{(k)} = \text{const} \tag{1.2.11}$$

Instead of considering different subsystems we can equally as well consider different phases. On the basis of the criterion that all extensive variables in the fundamental equation must be additive invariants (i.e. the principle of additive invariance) Gibbs formed the equation

$$\mathrm{d}S = (1/\theta)\mathrm{d}E + (p/\theta)\mathrm{d}V - (\mu_i/\theta)\mathrm{d}\boldsymbol{m}_i \tag{1.2.12}$$

which bears his name. In the Gibbs equation (1.2.12), the variables m_i are the masses. The conjugate intensive variables are defined by

$$(\partial_{\mathrm{m}_i} S)_{E,V,m_j} = -\mu_i/\theta \tag{1.2.13}$$

where μ_i are the chemical potentials or the partial specific Gibbs free energies of the components. In the absence of chemical reactions, the principle of additive invariance requires

$$\sum_k m_i^{(k)} = \text{const} \tag{1.2.14}$$

whereas the conservation of total mass

$$\sum_i \sum_k m_i^{(k)} = \text{const} \tag{1.2.15}$$

is always valid since the system is closed.

In open systems we will need a generalisation of the Gibbs equation since other parameters, besides the volume and masses of the components, are needed to define the state of the system (cf. chapter 5). The parameters, which do not enter into the classic equilibrium theory, are known as 'pseudo-thermodynamic' variables (Tisza, 1966). These variables do not obey the principle of additive invariance but rather vanish in the state of equilibrium. Examples of pseudo-thermodynamic variables are the electric and magnetic field strengths.

The use of composite systems in Gibbsian thermodynamics differs from the way they are employed in Carathéodory's theory. Composite systems, together with the thermodynamic process of the redistribution of additive invariants, are used to determine the conditions of equilibrium. This is a later development in Gibbsian thermodynamics. Gibbs did not use composite systems; rather, he based all his criteria of equilibrium (including that of stability) on the curvature of the primitive surface in the Gibbs space. The motivation for the use of composite systems in Gibbsian thermodynamics is the same as in Carathéodory's theory. It comes from the apparent paradox that if the isolated system is in equilibrium then the entropy can no longer increase. Alternatively, if the isolated system is not in equilibrium then the entropy is not definable. Whether or not an entropy can be defined for a non-equilibrium state is not a matter for equilibrium thermodynamics. In order to evade this problem, Gibbsian thermodynamics introduces the concept of a composite system.

Consider that the isolated system is divided into two subsystems by means of an adiabatic partition. Each of the separate subsystems is in a state of equilibrium. When the constraint, i.e. the adiabatic partition, is removed, a process is initiated in which there is a redistribution of additive invariants. The isolated system is in equilibrium both before and after the removal of the constraint. Thus, we consider the difference in entropy between two equilibrium states; that is, a more and a less constrained state of equilibrium. The resolution of the paradox in terms of composite systems has been discussed by Planck (1934, 1935) and Ehrenfest-Afanassjewa and de Haas-Lorentz (1935).

Now consider a specific application of composite systems in Gibbsian thermodynamics. Suppose that a rigid, adiabatic partition divides an isolated system into two subsystems, each of which is in equilibrium. When the constraints that the partition be adiabatic and rigid are released, i.e. it is replaced by a movable, diathermal partition, we have a redistribution of additive invariations such that

$$E^{(1)} + E^{(2)} = \text{const} \tag{1.2.16}$$

$$V^{(1)} + V^{(2)} = \text{const} \tag{1.2.17}$$

The Gibbs equation for the composite system is

$$\mathrm{d}S = (1/\theta_1)\mathrm{d}E_1 + (1/\theta_2)\mathrm{d}E_2 + (p_1/\theta_1)\mathrm{d}V_1 + (p_2/\theta_2)\mathrm{d}V_2 \tag{1.2.18}$$

since the diathermal partition is still restrictive to exchanges of matter. Introducing the constraints (1.2.16) and (1.2.17) into equation (1.2.18) leads to

$$\mathrm{d}S = (1/\theta_1 - 1/\theta_2)\mathrm{d}E_1 + (p_1/\theta_1 - p_2/\theta_2)\mathrm{d}V_1 \tag{1.2.19}$$

At equilibrium, the entropy is stationary with respect to all independent changes in the extensive variables. This implies that at equilibrium we must have

$$\theta_1 = \theta_2 \tag{1.2.20}$$

which is the condition of thermal equilibrium and

$$p_1 = p_2 \tag{1.2.21}$$

the condition of mechanical equilibrium.

Although both Carathéodory's theory and Gibbsian thermodynamics use composite systems to determine the conditions of equilibrium, Gibbsian thermodynamics imposes the further requirement that the extensive variables must be additive invariants. In fact, Carathéodory's theory does not even make the distinction between extensive and intensive variables. This is apparent when the equilibrium conditions (1.1.1) and (1.1.2) are compared to (1.2.20) and (1.2.21). Furthermore, one must bear in mind the distinction between processes that occur in the thermodynamic system and the 'conceptual' process of relaxing a constraint. It is preferable to call the relaxation of a thermodynamic constraint an 'operation' (Tisza, 1966) rather than an actual thermodynamic process.

1.3 Geometry of the Gibbs space

According to the basic tenets of Gibbsian thermodynamics, all the thermodynamic properties of the system are supposed to be contained in the fundamental equation. This equation can be represented as a surface in Gibbs space; it is called the primitive surface. From an analysis of the curvature of the primitive surface, Gibbs (1902) obtained his thermodynamic criteria of stability.

The entropy representation is not unique since, by definition

$$(\partial_E S)_V > 0 \tag{1.3.1}$$

the entropy constitutes a continuous, unique and differentiable function of the energy, the fundamental equation (1.2.8) can be solved for the internal energy. We then obtain

$$E = E(S, V) \tag{1.3.2}$$

The transformation from (1.2.8) to (1.3.2) amounts to a rotation in the Gibbs space. In Gibbsian thermodynamics, the entropy and energy representations can be used interchangeably since they both contain the same amount of thermodynamic information. This duality breaks down in non-equilibrium thermodynamics (cf. chapter 5). Even in equilibrium thermodynamics there are particular situations in which one representation is preferred to the other. For example, in equilibrium stability analysis, it is the energy and not the entropy representation which relates the stability criteria to directly measurable quantities.

Applying Euler's relation to the first order homogeneous equation (1.3.2) we obtain

$$E = \partial_{a_i} E a_i = X_i a_i \tag{1.3.3}$$

where $\{a_i\}$ stands for the complete set of extensive variables whose conjugate intensive variables belong to the set $\{X_i\}$. The equations of state are obtained through differentiation of the fundamental relation

$$\partial_{a_i} E = X_i \qquad \text{or} \qquad X_i = X_i(\{a_i\}) \tag{1.3.4}$$

They are interrelated through the integrability conditions, cf. condition (1.1.11),

$$\partial_{a_j} X_i = \partial_{a_i} X_j \tag{1.3.5}$$

These relations are commonly referred to as the Maxwell relations. If we differentiate the fundamental equation (1.3.3) without taking into consideration the specified functional dependencies and subtract the Gibbs equation, written in the form

$$\mathrm{d}E = X_i \mathrm{d}a_i \tag{1.3.6}$$

we come out with

$$a_i \mathrm{d}X_i = 0 \tag{1.3.7}$$

which is known as the Gibbs – Duhem relation. It is precisely what is needed to preserve the first order homogeneous property of the internal energy or, for that matter, the entropy. The Gibbs – Duhem relation is a statement to the effect that not all the intensive variables are independent. In fact, it is a relation among the intensive variables in differential form. In regard to the equations of state, we observe that the independent extensive variables can be eliminated to give a relation among the intensive variables. The same purpose would be served if the Gibbs – Duhem equation were integrable. A knowledge of all but one of the equations of state makes it possible to integrate the Gibbs – Duhem relation, thereby recovering the remaining equation of state. The process of integration introduces an arbitrary constant of integration so that this procedure can be used to obtain the fundamental equation, apart from an undetermined constant.

One connection between Gibbsian thermodynamics and geometry is furnished by the Legendre transformation. Recall that in Carathéodory's theory, the entropy is obtained through the integration of a Pfaffian differential equation and it is represented by a one parameter family of surfaces in the thermodynamic space. However, Carathéodory's theory is insensitive to which thermodynamic space is chosen, so that there is no way to obtain the characteristic potentials that are associated with particular thermodynamic spaces. Gibbs is to be accredited with the development of a transformation theory, that interchanges the roles between extensive and intensive variables. In this way, other spaces are introduced, each with its own characteristic potential. This displays the fundamental conceptual difference between Carathéodory theory which is aimed at a mathematical formulation of the entropy and Gibbs's theory which makes practical use of the entropy in the characterisation of thermodynamic systems.

The Legendre transform makes it possible to switch from an extensive variable to its conjugate intensive variable without any loss of information that is contained in the fundamental equation. In other words, the transformation is one-to-one; after having performed the transformation, we can apply the inverse transformation so as to recover the original equation. If the Legendre transforms are applied to the entropy, the resulting functions are known as

Massieu functions. However, the most useful thermodynamic potentials are obtained when the Legendre transformation is applied to the internal energy. This is another instance in which the energy representation is to be preferred to the entropy representation. The thermodynamic potentials that are obtained in the energy representation can be applied directly to the analysis of thermodynamic processes.

For simplicity, the Legendre transform of a single variable will be considered since it can be given a very simple geometrical interpretation in the plane. A sufficiently smooth curve can be represented either as a geometrical locus of points (x, y) or as the envelope of a family of tangents to the curve. The duality between the representation of a curve as a family of tangents and the representation of a curve as a geometrical locus of points is somewhat analogous to the duality that exists between extensive and intensive variables. In the x, y plane, we have the curve

$$y = y(x) \tag{1.3.8}$$

A new variable X is introduced by the transformation

$$X = \mathrm{d}_x y \tag{1.3.9}$$

Provided that

$$\mathrm{d}_x^2 y \neq 0 \tag{1.3.10}$$

we can solve for $x = x(X)$. If there were several independent variables to be transformed then the Hessian (i.e. the determinant formed from the second partial derivatives of y) must be different from zero. This guarantees the independence of the variables to be transformed.

The Legendre transform of $y(x)$ is defined as

$$Y(X) = Xx - y(x) \tag{1.3.11}$$

In the x, y plane, $Y(X)$ represents the intercept of the line on the ordinate axis and X is its slope. The functional dependency in (1.3.11) can be displayed through differentiation,

$$\mathrm{d}Y = X\mathrm{d}x + x\mathrm{d}X - \mathrm{d}y = x\mathrm{d}X \tag{1.3.12}$$

from which it follows that

$$x = \mathrm{d}_X Y \tag{1.3.13}$$

Now, in the case that

$$\mathrm{d}_X^2\, Y \neq 0 \tag{1.3.14}$$

we can solve equation (1.3.13) for $X = X(x)$. This allows us to eliminate X in equation (1.3.11) and we recover our original equation (1.3.8). The Legendre transform is, therefore, completely symmetrical. The same transformation that leads from the old to the new variable also leads back from the new to the old variable. In this sense, the Legendre transform is said to preserve all the thermodynamic information contained in the fundamental equation.

In Gibbs's theory, the equilibrium stability criteria are derived from the second variation of either the internal energy or the entropy, depending on the representation used. The second variations of these thermodynamic potentials are quadratic forms in the variations of the extensive variable from their equilibrium values. From an analogy with the theory of curvature in Euclidean space, it would seem reasonable to suppose that the quadratic forms should be related to the curvature of the primitive surface in Gibbs space. There is, however, the annoying point that the theory of curvature involves a metric which is expressed in terms of a line element. The notion of a metric is lacking in the geometrical structure of the Gibbs space.

Stability, in Gibbs's theory, is related to the positive definite or negative definite forms of the second variations of the internal energy and entropy, respectively. As has already been mentioned, the energy representation is better adapted for determining the equilibrium stability criteria. The stability criteria are associated with physically measurable quantities when the quadratic form is reduced to its canonical diagonal form. However, the reduction to diagonal form cannot be performed by the usual eigenvalue methods of metric definable spaces. Since a metric does not exist in Gibbs space, the reduction to diagonal form has to be carried out by the method of 'completing the square' (Tisza, 1966).

In order to derive the thermodynamic stability criteria, the internal energy is developed in a Taylor series expansion about the state of equilibrium. In the case of a single component system, the stability criteria are particularly transparent. In this case, if the internal energy is expanded in terms of all three independent variables, S, V and m, the determinant of the coefficient matrix of the quadratic form will vanish. It is soon seen why this is so.

We therefore divide the internal energy by the mass and obtain the specific internal energy ε,

$$\varepsilon = \varepsilon(\eta, \rho^{-1}) \tag{1.3.15}$$

where η is the specific entropy and ρ is the density. For a compact notation, we represent the independent variables by α_i and expand (1.3.15) about the equilibrium state. For small variations from the equilibrium state, it is sufficient to retain only first and second order terms in the series expansion

$$\varepsilon(\alpha_1, \alpha_2) = \varepsilon_0 + X_i \xi_i + \tfrac{1}{2} E_{ij} \xi_i \xi_j \tag{1.3.16}$$

where ξ_i are the deviations in the extensive variables from their equilibrium values, α_{i0}, namely

$$\xi_i = \alpha_i - \alpha_{i0} \tag{1.3.17}$$

The forces X_i are still defined by (1.3.4) or equivalently

$$X_i = \partial_{\alpha_i} \varepsilon \tag{1.3.18}$$

and

$$E_{ij} = \partial_{\alpha_i} \partial_{\alpha_j} \varepsilon \tag{1.3.19}$$

The matrix of differential coefficients, $\mathbf{E}$, is referred to as the 'stiffness' matrix (Tisza, 1966). The reason for this name is that its elements are related to the force variations by

$$\chi_i = E_{ij} \xi_j \tag{1.3.20}$$

where the force variations are defined by

$$\chi_i = X_i - X_{i0} \tag{1.3.21}$$

If we had expanded the specific energy in all the independent variables the determinant of the linear equations in (1.3.20) would have vanished. This is due to the fact that the Gibbs – Duhem relation (1.3.7) constitutes a linear relation among the three equations.

We now have to introduce the fact that the system is evaluated at equilibrium. This is achieved by coupling the thermodynamic system to another system whose specific internal energy ε' is also developed in a Taylor series

$$\varepsilon'(\alpha_1', \alpha_2') = \varepsilon_0' + X_i' \xi_i' + \tfrac{1}{2} E_{ij}' \xi_i' \xi_j' \tag{1.3.22}$$

Therefore, the total change in the specific internal energy is

$$\Delta\varepsilon + \Delta\varepsilon' = \chi_i \xi_i + \tfrac{1}{2} E_{ij} \xi_i \xi_j + \chi_i' \xi_i' + \tfrac{1}{2} E_{ij}' \xi_i' \xi_j' \tag{1.3.23}$$

We now consider the limiting case where the mass of the primed system is infinitely greater than the unprimed system. Since $E_{ij} = E_{ij}'$, the last member in (1.3.23) is negligible in comparison to the other terms. In other words, the prime system behaves as a reservoir whose coupling to the system can be adequately described by the unperturbed forces X_i'.

The condition that the system should be in the state of equilibrium is introduced by the principle of additive invariance. It can be expressed as

$$\xi_i + \xi_i' = 0 \tag{1.3.24}$$

Introducing conditions (1.3.24) into (1.3.23) leads to

$$\Delta\varepsilon + \Delta\varepsilon' = (X_i - X_i')\xi_i + \tfrac{1}{2}\delta^2\varepsilon \tag{1.3.25}$$

where

$$\delta^2\varepsilon = E_{ij} \xi_i \xi_j \tag{1.3.26}$$

The equilibrium conditions are

$$X_i = X_i' \tag{1.3.27}$$

which makes the total energy an extremum provided (1.3.26) does not vanish. The Gibbs stability criterion can now be formulated as: in an isolated system the internal energy tends to a minimum at constant entropy. An alternative version of the stability criterion is: in an isolated system, the entropy tends to a maximum at constant internal energy. It is apparent that we are dealing with the first of the two criteria of Gibbs; the constancy of the entropy is expressed by (1.3.24). It is now necessary to show that (1.3.26) is positive definite in all stable equlibrium systems.

The stability conditions can be related to physically measurable quantities by casting the quadratic form (1.3.26) in the canonical diagonal form, provided of course that the determinant of the quadratic form is non-singular, namely

$$D_2 = |E_{ij}| \neq 0 \tag{1.3.28}$$

Since we know that the Gibbs space is one in which volumes and not lengths are measurable, we require the linear (affine) transformation to be unimodular (of determinant equal to unity), for only then will the volume be invariant under the transformation (Tisza, 1966). The method consists of completing the square and we finally arrive at the canonical diagonal form

$$\delta^2 \varepsilon = \lambda_k \phi_k^2 \tag{1.3.29}$$

Stability requires that each of the coefficients in (1.3.29) is to be positive. The determinant of the coefficient matrix is simply the product of the coefficients, namely

$$D_2 = \lambda_1 \lambda_2 \tag{1.3.30}$$

where

$$D_2 = \begin{vmatrix} E_{11} & E_{12} \\ E_{21} & E_{22} \end{vmatrix} = \frac{\partial(\theta, -p)}{\partial(\eta, \rho^{-1})} \tag{1.3.31}$$

with $E_{12} = E_{21}$. From (1.3.30), we have

$$\lambda_1 = D_1 \qquad \text{and} \qquad \lambda_2 = D_2/D_1 \tag{1.3.32}$$

The first criterion of stability is

$$\lambda_1 = (\partial_\eta \theta)_\rho = \theta / C_V > 0 \tag{1.3.33}$$

where C_V is the specific heat capacity at constant volume. The second criterion of stability can be written as

$$\lambda_2 = \frac{\partial(\theta, -p)}{\partial(\eta, \rho^{-1})} \cdot \frac{\partial(\eta, \rho^{-1})}{\partial(\theta, \rho^{-1})} = -(\partial_{\rho^{-1}} p)_\theta = \rho / K_\theta > 0 \tag{1.3.34}$$

where K_θ is the coefficient of isothermal compressibility. Therefore, the conditions

$$C_V > 0 \qquad \text{and} \qquad K_\theta > 0 \tag{1.3.35}$$

are both necessary and sufficient to ensure that the second variation of the specific internal energy (1.3.26) is positive definite. We have thus established the positive definiteness of the quadratic form (1.3.26) as a necessary and sufficient criterion of stability on the basis that the conditions (1.3.35) must always be fulfilled in all stable systems.

In the multi-dimensional case, the series expansion can be centred on any point on the primitive surface. The points on the primitive surface fall into three classes: (1) elliptic points for which all $\lambda_k > 0$, (2) parabolic points where all $\lambda_k > 0$ with at least one $\lambda_k = 0$, and (3) hyperbolic points with at least one $\lambda_k < 0$. If the point is elliptic, then the equilibrium is stable. Alternatively, a hyperbolic point implies instability. The existence of a parabolic point indicates that the equilibrium is metastable.

1.4 Equilibrium extremum principles

Although the two versions of the Gibbs stability criterion are equivalent, there is an interesting conceptual difference between the two forms. In order to maintain the entropy constant in the energy minimum principle, we have seen that it is necessary to couple the system to an external body, i.e. a Carnot cycle. At constant entropy, the internal energy behaves as any other potential energy of classical mechanics. This is why the stability conditions can be related to directly measurable quantities. In the entropy representation, the Gibbs stability criterion is that in an isolated system the entropy tends to a maximum at constant energy. If the isolated system has been left alone for a sufficiently long period of time, we would expect that equilibrium will have been established and the entropy can no longer increase, since it is already at its maximum value. How then is it possible to apply the entropy criterion?

This point is obfuscate in Gibbs's presentation. We now know that the answer lies with the concept of a composite system. When this is used in conjunction with the principle of additive invariance, we have a means by which we can explicitly introduce the fact that the system is indeed isolated. This usage of composite system differs from that of Carathéodory's theory (cf. section 1.1). Moreover, we have seen that the type of equilibrium that we wish to establish is reflected in the properties of the partitions. In addition, there is an order in which the different types of equilibrium can be established. Since there can be energy flow in the absence of matter flow, the conditions of thermal equilibrium must necessarily precede those of chemical equilibrium.

If we consider an isolated system which is divided into a number of subsystems by adiabatic partitions, then when we relax this constraint, say by replacing the adiabatic partitions by diathermal partitions, a process will be initiated that redistributes the extensive variables among the subsystems. The notion that the system is isolated is introduced by the condition that the extensive variables must satisfy the principle of additive invariance. These conditions are analogous to the kinematic constraints in classical mechanical variational principles. That is, in order to find an equilibrium position of a mechanical system, the potential energy is required to be a minimum subject to the kinematical constraints that are ordinarily imposed by the particular geometrical configuration. The constraints imposed by the principle of additive invariance are holonomic (i.e. they are relations among the variables themselves instead of their differentials) and sclereonomic (i.e. they do not involve time). It will also be recalled that an additional attribute of composite systems in Gibbsian thermodynamics is that it is only necessary to compare the entropy of two equilibrium states—a more and a less constrained state of equilibrium. In this way, the problem of defining the entropy of a non-equilibrium state is avoided.

The entropy maximum principle is an equilibrium differential or extremum principle. It is used to determine both the conditions of equilibrium and the stability of the equilibrium. If we treat a composite system in which the adiabatic partitions have been suddenly replaced by diathermal partitions, we obtain the condition of thermal equilibrium from the extremum principle

$$S = S(E) = \sum_{k} S^{(k)}(E^{(k)}) = \max \tag{1.4.1}$$

The extremum principle can be used to establish other types of equilibrium when we use more sophisticated partitions. However, we will content ourselves with the determination of the condition of thermal equilibrium and the stability of the equilibrium, both of which can be derived from (1.4.1).

On account of the relaxation of the adiabatic constraint, (1.4.1) is not a free extremum principle but rather a 'conditioned' one since it is subject to the constraint (1.2.11). This constraint can be introduced explicitly into the extremum principle (1.4.1) by the method of Lagrange's undetermined multiplier. This method permits a free variation of all the extensive variables. Alternatively, we can eliminate one of the internal energy variations through (1.2.11) but this has the disadvantage of making the extremum principle unsymmetrical. The method of the Lagrange undetermined multiplier preserves the symmetry of the extremum principle.

If we multiply the constraint (1.2.11) by the Lagrange undetermined multiplier λ and add it to the extremum principle (1.4.1) we obtain

$$S_I = S + \lambda \sum_k E^{(k)} = \max \tag{1.4.2}$$

The new function S_I can be interpreted as the entropy of the isolated system plus that which is needed to maintain the constraint (1.2.11). The condition of stationarity is

$$\sum_k \{\partial_{E^{(k)}} S^{(k)} + \lambda\} \, \delta E^{(k)} = 0 \tag{1.4.3}$$

where we have used the property that the entropies of the subsystems are additive. Let us also observe that $\delta S_I = \delta S$ since all we have done is to add zero to the stationarity condition (1.4.3), cf. constraint (1.2.11). The variations $\delta E^{(k)}$ are now all independent and in order to satisfy the stationary condition (1.4.3) identically we must put

$$\partial_{E^{(k)}} S^{(k)} = -\lambda \tag{1.4.4}$$

Condition (1.4.4) must be true for each and every subsystem k; the Lagrange multiplier is identified as the negative of the inverse of the absolute temperature. At thermal equilibrium, the absolute temperature must be constant throughout the isolated system. This result has already been obtained by another method in (1.2.20). However, the extremum principle offers us a bonus in that the second variation of (1.4.1) provides the criterion of thermal stability, namely

$$\partial^2_E S = -(\theta^2 C_V)^{-1} < 0 \tag{1.4.5}$$

which will be recognised as the first equilibrium stability criterion that we have obtained in the energy representation, (1.3.35). Inequality (1.4.5) states that the entropy surface is convex, with respect to the energy axis in the Gilbbs space, in all stable systems that are in thermal equilibrium.

1.5 Principle of maximum work

Section 1.1 discussed Carathéodory's theorem, which expresses the equivalence between the inaccessibility of states from those that lie along $dQ = 0$ and the existence of an integrating denominator. This is to say that the inaccessible points lie on paths which are irreversible and cannot be reached from adiabatics unless work is done. It is perhaps this concept of work that is associated with irreversibility which may ultimately lead to a theory of non-equilibrium thermodynamics based on Carathéodory's theory.

The connection between work and irreversibility can also be seen in Gibbsian thermodynamics. Specifically, we ask for the maximum work that a body can do on an external object which is adiabatically enclosed (Landau and Lifshitz, 1958). If we were to consider the system as comprised of the body and the object alone the answer would be simple: the work done by the body is equal to the decrease in its internal energy. However, if we introduce a medium, which together with the body and object form an isolated system, then this is no longer true and it then makes sense to ask about the maximum work that the body can do on the object. Alternatively, if work is done by the object on the body, we would want to determine the minimum work that can be done.

The body and the medium can exchange heat and work since we put them at two different temperatures θ and θ_0 and at two different pressures p and p_0. The subscript zero refers to the properties of the medium; for simplicity, we consider only compressional work. As the body performs work on the object, it makes a transition from one state to another. During the transition, the body may exchange heat and work with the medium. From the first law of thermodynamics we have

$$\Delta E = W + p_0 \Delta V_0 + \Delta Q \tag{1.5.1}$$

where ΔE is the total change in the internal energy of the body, which may be a finite instead of an infinitesimal change, and W is the work done on the body whose initial state is kept fixed so that W depends only on the final state of the transition. The second term in (1.5.1) is the compressional work done by the medium on the body and the last term is the heat gained from the medium.

Owing to the large size of the medium, its temperature and pressure are assumed to be constant. Moreover, we suppose that the process by which heat is given up to the body occurs reversibly, i.e. $\Delta Q = -\theta_0 \Delta S_0$. Thus, the principle of conservation of energy can be expressed as

$$\Delta E = W + p_0 \Delta V_0 - \theta_0 \Delta S_0 \tag{1.5.2}$$

Since the entire system is isolated, the total entropy must increase or at most remain the same,

$$\Delta S + \Delta S_0 \geq 0 \tag{1.5.3}$$

while the total volume of the system satisfies the principle of additive invariance

$$\Delta V + \Delta V_0 = 0 \tag{1.5.4}$$

Introducing conditions (1.5.2) and (1.5.3) into equation (1.5.2) leads to the inequality

$$W \geq \Delta E - \theta_0 \Delta S + p_0 \Delta V \tag{1.5.5}$$

where the equality holds in a reversible process. It will now be appreciated that, in a reversible process, the object does a minimum amount of work on the body, while in the reverse process the body will perform a maximum amount of work on the object. Hence

$$|W|_{max} = -\Delta(E - \theta_0 S + p_0 V) \tag{1.5.6}$$

What we have done essentially is to interchange initial and final states. We can infer from (1.5.6) that if the transition is not reversible then some of the work is lost in 'other' processes. This work cannot be recovered. Inequalities such as (1.5.6) can provide no information concerning the processes which dissipate energy, since we would have to know the rate of working of the object. This will be discussed in full detail in chapter 5. Let it suffice here to say that there is a very deep relationship between work and irreversibility.

Equilibrium thermodynamics does, however, allow us to associate the minimum work done on the body with its internal processes. The Gibbs equation (1.3.6) can be written in the form

$$\Delta E = \theta \Delta S - p \Delta V + X_i' \Delta a_i' \tag{1.5.7}$$

where a_i' are those variables, besides the volume, which are necessary to define the state of the body. Introducing (1.5.7) into (1.5.5) results in

$$W \geq (\theta - \theta_0)\Delta S - (p - p_0)\Delta V + X_i' \Delta a_i' \tag{1.5.8}$$

It is important to observe that if the body were defined only in terms of θ, p or V, then the constancy of these variables would imply that no process would occur. Therefore, only in the case that there are other extensive variables, which are required to define the body while obeying the principle of additive invariance, can we have a process occurring in an isolated system that is both in thermal and mechanical equilibrium. The additional extensive variables in the Gibbs equation (1.5.7) must necessarily be related to internal processes that are occurring in the body. Moreover, it is necessary that the body is not in equilibrium with the medium with respect to these internal processes.

Under conditions of constant temperature and pressure, inequality (1.5.8) reduces to

$$W \geq (\Delta G)_{\theta, p} \tag{1.5.9}$$

where G is the Gibbs free energy. For a spontaneous, irreversible process which brings the system to equilibrium

$$(\Delta G)_{\theta, p} \leq 0 \qquad (W = 0) \tag{1.5.10}$$

Under conditions of constant temperature and pressure, in which there is no work done, the Gibbs free energy cannot increase and it will reach a minimum at equilibrium. In other words, the equilibrium values of the unconstrained parameters in the body which is in contact with a temperature and a pressure reservoir (i.e. the medium) minimises the Gibbs function at constant temperature and pressure.

This concludes the summary of the formulations of equilibrium thermodynamics. We now proceed beyond the limits of equilibrium thermodynamics and discuss the classic and more recent theories of linear and nonlinear thermodynamics. During the analyses there will be occasion to return to the equilibrium formulations of thermodynamics since they provide the basis upon which a theory of non-equilibrium thermodynamics can be constructed.

References

Blaschke, W. (1923). *Vorlesungen über Differential Geometrie,* vol. II, Affine Differential Geometrie, ch. 4, Springer-Verlag, Berlin.

Born, M. (1921). *Phys. Z.,* **22,** 218, 249, 282.

Born, M. (1949). *Natural Philosophy of Cause and Chance,* Clarendon Press, Oxford.

Buchdahl, H. A. (1949). *Am. J. Phys.,* **17,** 41, 44, 212.

Buchdahl, H. A. (1954). *Am. J. Phys.,* **22,** 182.

Buchdahl, H. A. (1955). *Am. J. Phys.,* **23,** 65.

Carathéodory, C. (1909). *Math. Annln,* **67,** 335.

Carathéodory, C. (1925). *Sber. preuss. Akad. Wiss.,* 39.

Chandrasekhar, S. (1939). *Introduction to the Study of Stellar Structure,* ch. 1, Chicago University Press, Chicago.

Clausius, R. (1850). *Pogg. Ann. Phys.,* **79,** 378, 500.

Eckart, C. (1940). The thermodynamics of irreversible processes II. Fluid mixtures. *Phys. Rev.,* **58,** 269–275.

Ehrenfest, P. (1911). *Z. phys. Chem.,* **77,** 227.

Ehrenfest-Afanassjewa, T. (1925). *Z. Phys,* **33,** 933; (1926), **34,** 638.

Ehrenfest-Afanassjewa, R. and de Haas-Lorentz, G. L. (1935). *Physica,* **2,** 743.

Eisenschitz, R. (1955). *Sci. Prog.,* **43,** 246.

Gibbs, J. W. (1902). *Collected Works,* Scribner, New York.

Goursat, E. (1922). *Leçons sur le problème de Pfaff,* Hermann, Paris.

Landau, L. D. and Lifshitz, E. M. (1958). *Statistical Physics,* Pergamon Press, Oxford.

Landé, A. (1926). *Handbuch der Physik,* vol. 9, ch. IV, Springer-Verlag.

Landsberg, P. T. (1956). Foundations of thermodynamics. *Rev. mod. Phys.,* **28,** 363–392.

Landsberg, P. T. (1961). *Thermodynamics with Quantum Statistical Illustrations,* Interscience, New York.

Margenau, H. and Murphy, G. M. (1943). *The Mathematics of Physics and Chemistry,* Van Nostrand, New York.

Planck, M. (1934). *Annln Phys.* **19,** 759.

Planck, M. (1935). *Physica,* **2,** 1029.

Planck, M. (1954). *Treatise on Thermodynamics,* Dover, New York.

Schouten, J. A. and van der Kulk, W. (1949). *Pfaff's Problem and its Generalizations,* Oxford University Press, New York.

Thomson, W. (1848). *Edinburgh Trans.,* **16,** pt V, 54.

Tisza, L. (1966). *Generalized Thermodynamics,* MIT Press, Cambridge, Mass.

2 Classical Non-equilibrium Thermodynamics

In this chapter the treatise of non-equilibrium thermodynamics is begun. Although the groundwork was laid by the classic investigations of Kelvin, Maxwell, Rayleigh, Thomson and others, in the latter half of the nineteenth century, the classical theory of non-equilibrium thermodynamics did not emerge until the first half of this century when Onsager (1931) firmly laid down its basic principles. In Onsager's work, we note a confluence of ideas that were borrowed from both macroscopic and microscopic physics. His endeavours culminated in the derivation of what are now commonly referred to as the 'Onsager reciprocal relations'. While the motivation for their derivation is to be found in Kelvin's earlier work on thermoelectric phenomena, the derivation of the Onsager reciprocal relations is based upon the principle of microscopic reversibility.

The years following Onsager's original inquiries into non-equilibrium thermodynamics witnessed the growth of a vast literature on the interpretation and application of the Onsager reciprocal relations. It eventually grew into the classical theory of non-equilibrium thermodynamics as we know it today. In classical non-equilibrium thermodynamics, the Onsager reciprocal relations are attributed with a universal validity and applications have ranged from physics (de Groot and Mazur, 1962) to biology (Katchalsky and Curran, 1965). Although it was clearly observed that non-equilibrium thermodynamics overlaps with other branches of macroscopic physics, scepticism was voiced that non-equilibrium thermodynamics could not add anything new to what was already known (Wei, 1966). Criticisms of this nature are, in part, valid. However, they do not stem from Onsager's original formulation but rather are due to the indiscriminate applications and vague interpretations of the *class* of reciprocal relations that bears Onsager's name. In order to appreciate these remarks, it is necessary to trace the paths that lead to and from Onsager's formulation of non-equilibrium thermodynamics.

2.1 The predecessors of Onsager

Reciprocal relations were known for quite some time prior to Onsager (1931).

Although reciprocal relations were considered almost a truism by investigators of the second half of the nineteenth century, they were nevertheless destined to dominate the scene of twentieth century non-equilibrium thermodynamics. As is often the case, much is lost in the translation of original formulations. It is therefore worth while to discuss the reciprocal relations as they were originally construed.

Consider the mutual effect that one body has upon another when both are fixed in space by the generalised coordinates q_1 and q_2. A force F_1 causes a change δq_1 in its conjugate generalised coordinate q_1. The magnitude of the change may also depend upon the other coordinate q_2. Suppose that q_2 changes by an amount δq_2 and that a force variation δF_1 is required to keep q_1 at its original value. Then if W is the scalar potential,

$$F_1 = -\partial_{q_1} W \tag{2.1.1}$$

and

$$F_1 + \delta F_1 = -\partial_{q_1} W - \partial_{q_2} \partial_{q_1} W \delta q_2 \tag{2.1.2}$$

where

$$\delta F_1 = -\partial_{q_2} \partial_{q_1} W \delta q_2 \tag{2.1.3}$$

Now consider the opposite situation in which the force F_2, conjugate to q_2, produces a change in q_2 which also causes a change in q_1. If δF_2 is the amount necessary to keep q_2 constant then

$$F_2 = -\partial_{q_2} W \tag{2.1.4}$$

and

$$F_2 + \delta F_2 = -\partial_{q_2} W - \partial_{q_1} \partial_{q_2} W \delta q_1 \tag{2.1.5}$$

where

$$\delta F_2 = -\partial_{q_1} \partial_{q_2} W \delta q_1 \tag{2.1.6}$$

Then from (2.1.3) and (2.1.6) we obtain the 'reciprocal relation'

$$(\partial_{q_2} F_1)_{q_1} = (\partial_{q_1} F_2)_{q_2} \tag{2.1.7}$$

which, expressed in words, states that the change in F_1 due to an increase in q_2 at q_1 constant is the same as the change in F_2 due to an increase in q_1 at q_2 constant. In other words, if F_1 depends on q_2 then F_2 will depend on q_1 in the same manner. Reciprocal relations of the form (2.1.7) pervade the literature of the latter half of the nineteenth century. The mechanical formulation was given by Rayleigh in his *Theory of Sound* (cf. section 7.5), whereas their formulation in thermodynamics is to be found in Maxwell's *Theory of Heat*.

One may realise at this point that (2.1.7) is none other than the exactness condition for the scalar potential W. There are other types of force that arise from friction or viscosity. These forces are proportional to the velocities instead of the displacements. Then, if exactness conditions of the form (2.1.7), with the generalised coordinates replaced by the generalised velocities, are satisfied, it would imply that there exists a potential for these forces. This scalar potential was first discovered by Rayleigh (1873, 1877) and it is called the 'dissipation function'.

It would be a trivial matter indeed if the dissipation function was the time derivative of a scalar potential such as W, since all that would be necessary is to replace W by its time derivative and the generalised coordinates by the generalised velocities. This is, however, not the case since the forces derived from W are functions of the generalised coordinates and not the generalised velocities. Consequently, the dissipation function has a character that is uniquely different from the other scalar potentials of classical mechanics. In the same manner, we cannot expect the kinetic analogues of the thermodynamic Maxwell relations to be formed by simply taking the time derivative of the thermodynamic potentials. Additional concepts and information are required in order to make the transition from equilibrium to non-equilibrium thermodynamics and this is what the Onsager reciprocal relations are all about.

Reciprocal relations were found to apply to a vast range of diverse phenomena. Lord Kelvin (1854) showed that when a current flows along a bar with a non-uniform distribution of temperature, heat is also conducted. Then it follows from the reciprocal relation that an electromotive force will be developed in any conductor in which there is a flow of heat. Logical as the conclusion may appear, it will be appreciated that we are no longer dealing with the time derivative of a generalised coordinate but rather with material fluxes. This distinction was to cause a great deal of controversy in the interpretation of the Onsager reciprocal relations (Casimir, 1945; Truesdell, 1969). But since we are dealing with an historical survey of the reciprocal relations, it is interesting to see how such phenomena were treated before the advent of non-equilibrium thermodynamics.

Let us consider the approach of Thomson (1888) which attempts to apply dynamical methods to processes that involve the conversion of energy into heat. However, we must be careful to understand Thomson's approach in an historical context. In his day, it was the vogue to attempt the derivation of the second law from the principle of least action. It was, however, appreciated that the second law is derived from experience and it is not a purely dynamical principle. In addition, the first law or the principle of conservation of energy was considered to be a dynamical result rather than a dynamical method. So Thomson asked what degree of success would be achieved by using classical mechanics in the analyses involving transfers of heat and energy.

Thomson's method divides the kinetic energy into two parts, depending on whether the motion is that of a 'controllable' coordinate q_i or an 'uncontrollable' coordinate which we need not specify. The variations of the controllable coordinates determine the performance of work, such as variations in strain or electric displacement. In other words, they comprise the set of extensive thermodynamic and pseudo-thermodynamic variables that was discussed in chapter 1. The motion of these variables gives rise to a kinetic energy T_c which Helmholtz called the 'freie Energie' in his classic paper 'Die Thermodynamik chemischer Vorgänge'. Relative motion of the system components creates another type of kinetic energy T_u which is equated to the mean temperature of the system. This is the dynamical interpretation of temperature that is offered by the kinetic theory of gases and it provides the connection between heat and mechanics.

For a system with a single controllable coordinate, the conservation of energy is

$$\delta Q - F\delta q = \delta T_c \tag{2.1.8}$$

δQ is the quantity of heat that is transferred to the system and T_c is a second order homogeneous function of the velocity of the controllable coordinate,

$$2T_c = \dot{q}\mathrm{d}_{\dot{q}} T_c \tag{2.1.9}$$

Later, we shall appreciate that Helmholtz's interpretation of T_c and the functional form (2.1.9) brings out the intimate connection between the free energy and the dissipation function, cf. equation (3.3.35).

The variation of (2.1.9) is explicitly introduced into (2.1.8). Then using the equation of motion

$$\mathrm{d}_t \mathrm{d}_q T_c - \mathrm{d}_q T_u = F \tag{2.1.10}$$

we obtain

$$\delta Q = -\mathrm{d}_q T_u \delta q \tag{2.1.11}$$

T_u cannot be a function of the velocity of the controllable coordinate, otherwise we would have to know the velocity in order to measure the temperature. Thus, T_u can only be a function of q and we suppose this functional dependency to be given by $f(q)$. We can therefore write

$$\mathrm{d}_q T_u = f'(q)/f(q) \cdot T_u \tag{2.1.12}$$

Introducing (2.1.12) into (2.1.10) and differentiating with respect to the energy due to the motion of the uncontrollable coordinates (that is, the temperature), at constant values of the controllable coordinates, we obtain

$$\mathrm{d}_{T_u} F = -f'(q)/f(q) \tag{2.1.13}$$

In view of (2.1.12) we arrive at the expression

$$\delta Q = \theta(\mathrm{d}_\theta F)_q \delta q \tag{2.1.14}$$

where we have replaced T_u by the absolute temperature θ. Expression (2.1.14) can also be written as

$$[\mathrm{d}_q(Q/\theta)]_\theta = (\mathrm{d}_\theta F)_q \tag{2.1.15}$$

Calling $\delta Q/\theta$ the entropy, it will be appreciated that (2.1.15) is none other than a Maxwell relation (for example $F = p$ and $q = V$).

Thomson (1888) interpreted (2.1.14) as the heat δQ which must be supplied to the system in order to prevent its temperature from changing due to an increase δq in the controllable coordinate. In the particular case of Kelvin's thermoelectric effect, δQ is the quantity of heat that must be supplied to a unit volume of the conductor to prevent its temperature from changing when there is an increase in electricity δq. Then if I is the current,

$$\delta q = I\,\delta t \tag{2.1.16}$$

the expression for the 'entropy production' becomes

$$\frac{\delta}{\delta t}(Q/\theta) = (\mathrm{d}_\theta F)_q \cdot I \tag{2.1.17}$$

where F is the electromotive force that can be related to the temperature gradient.

According to Thomson's approach, the fluxes will always be the time derivative of the extensive or pseudo-thermodynamic variables. Results such as (2.1.14) and (2.1.17) were first derived by Kelvin in his *Dynamical Theory of Heat.* Since it makes no sense to ask about the reciprocal effect caused by an increase in $\delta\theta$ (that is, δQ is always that quantity of heat required to maintain the temperature constant), Kelvin had to make the additional assumption: 'The electromotive forces produced by inequalities of temperature in a circuit of different metals, and the thermal effects of electric currents circulating in it, are subject to the laws which would follow from the general principles of the thermodynamic theory of heat if there were no conduction of heat from one part of the circuit to another.' Moreover, it is interesting to note that neither Kelvin nor Thomson considered the reciprocal phenomena as irreversible effects that would be caused by friction or viscosity.

The foundations of the classical theory of non-equilibrium thermodynamics were constructed by Onsager (1931) who was motivated by Kelvin's thermoelectric phenomenon. Onsager wanted to derive the Kelvin reciprocal relation from 'recognised fundamental principles'. This he did not accomplish. However, Onsager did establish a *class* of reciprocal relations which has a necessary condition that the flux be the time derivative of an extensive or pseudo-thermodynamic variable. Our hindsight will allow us to appreciate that the class of reciprocal relations are applicable to the Kelvin – Thomson formulation but it is not applicable to those cases where the thermodynamic flux is an actual material flux.

Moreover, Thomson's method is an attempt to account for dissipative effects in terms of the controllable state variables. It will be appreciated at a later stage that this approach is very similar to the internal state variable formulation of dissipative phenomena in nonlinear thermodynamics (cf. sections 5.2, 9.2 and 9.5).

2.2 The statistical basis of Onsager's reciprocal relations

Although Kelvin's relations provided the motivation, the derivation of the Onsager reciprocal relations was based on an apparent analogy between the conditions of chemical equilibrium and the 'principle of detailed balance at equilibrium' (Fowler, 1929). However, as Fowler pointed out, a detailed balancing of elementary processes could be achieved through cyclical processes which would destroy the intended significance of the principle. Therefore, in order to use the principle, one has to group every elementary process with its reverse, just as one does to establish the conditions of chemical equilibrium by equating the velocities of forward and backward reactions. The proof of the Onsager reciprocal relations then took substance in the application of the corresponding microscopic principle, referred to by Tolman (1938) as the 'principle of equal frequency for reverse molecular processes at equilibrium' or its more familiar title of the 'principle of microscopic reversibility', to fluctuations that occur in an aged system.

Onsager's proof, which is based on the equilibrium theory of fluctuations, is reproduced in all the standard texts on non-equilibrium thermodynamics (de Groot, 1952; de Groot and Mazur, 1962; Landau and Lifshitz, 1958; Prigogine, 1967; Yourgrau, van der Merwe and Raw, 1966). The first critical analysis of the derivation and generality of the Onsager reciprocal relations was given by Casimir (1945).

The proof is based on the validity of the Einstein (1910) formula for the probability of a fluctuation. This formula relates the probability of a spontaneous fluctuation from equilibrium to the change in entropy. Specifically, one supposes that a phenomenological state of an adiabatically insulated system is described by the set of thermodynamic variables $\{\xi\}$. The ξ_i's are the deviations from their equilibrium values α_i^0 of the *extensive* variables α_i, namely

$$\xi_i = \alpha_i - \alpha_i^0 \tag{2.2.1}$$

The entropy change is

$$\Delta\eta = \eta - \eta_0 \tag{2.2.2}$$

where η_0 is the equilibrium value of the entropy. Since the entropy change cannot depend on the way in which the deviation from equilibrium has occurred, $\Delta\eta$ is required to have the form

$$\Delta\eta = -\tfrac{1}{2}S_{ij}\,\xi_i\xi_j \tag{2.2.3}$$

which is valid provided the deviations from equilibrium are small. Since the entropy at equilibrium is a maximum, the first variation in (2.2.3) vanishes leaving the quadratic expression. According to the equilibrium conditions of stability, the matrix of coefficients $\|S_{ij}\|$ must be positive semidefinite. Then, according to Einstein (1910), the probability of a spontaneous fluctuation from equilibrium will be given by

$$P(\{\xi\}) \propto \exp(\Delta\eta/k) \tag{2.2.4}$$

where k is Boltzmann's constant.

The Einstein formula (2.2.4) can be used to calculate average values of fluctuating quantities. The thermodynamic forces are defined by

$$X_i = -\partial_{\alpha_i}\eta \tag{2.2.5}$$

whose variations are

$$\chi_i = -\partial_{\xi_i}\Delta\eta = S_{ij}\xi_j \tag{2.2.6}$$

We see that the thermodynamic force driving the system to equilibrium is linear in the displacement from equilibrium. Simply by performing a Gaussian integration, it can be shown that

$$\langle\chi_i\xi_k\rangle = k\delta_{ik} \tag{2.2.7}$$

where δ_{ik} is the delta function of Kronecker. Relation (2.2.7) will be of use to us in the Onsager–Casimir derivation of a class of reciprocal relations, which we shall now discuss in some detail.

The (fluctuating) thermodynamic variables are supposed to be even functions of the molecular velocities. This implies that they should be invariant under a time reversal transformation. Then according to the Onsager–Casimir interpretation, the principle of microscopic reversibility at equilibrium requires

$$\langle \xi_k(t+\tau) \rangle = \langle \xi_k(t-\tau) \rangle \tag{2.2.8}$$

Some explanation of the average in (2.2.8) is required, since at first glance we would suppose it to vanish. Onsager (1931) tells us to calculate it in the following way: every time the complete set of fluctuating variables is found to be in the same known configuration $\{\xi(t)\}$ at time t, we note down the displacement ξ_k τ sec later. The experiment is to be repeated a great number of times and the average of all our results is given by the left-hand side of (2.2.8). In a certain sense, we have avoided defining a non-equilibrium average. Then (2.2.8) states that this average will be the same as the average of ξ_k calculated at τ sec prior to the configuration $\{\xi(t)\}$. At time $\tau = 0$ we have

$$\langle \xi_k(t) \rangle = 0 \tag{2.2.9}$$

which Onsager claims is a result of symmetry. We would like to think of it as an equilibrium average.

According to (2.2.8) we can expect, on the average, the past and future behaviour of the process to be the same. In other words, there is a symmetry in past and future.

If (2.2.8) is multiplied by $\xi_j(t)$ and the average taken over all possible values of the variables, we obtain

$$\langle \xi_j(t)\xi_k(t+\tau) \rangle = \langle \xi_j(t)\xi_k(t-\tau) \rangle \tag{2.2.10}$$

The (equilibrium) time correlation expressions are to be evaluated with the aid of the phenomenological equations describing the regression of the (fluctuating) thermodynamic variables

$$\dot{\xi}_k = L_{kj}\xi_j \tag{2.2.11}$$

A small point: these are the empirical laws which the fluctuations obey only on the average. It would be more correct to introduce angular brackets in (2.2.11) indicating a non-equilibrium average.

It would facilitate matters if we could utilise expression (2.2.7). To be able to do so we must transform equations (2.2.11) into the form

$$\dot{\xi}_k = \Lambda_{kj}\chi_j \tag{2.2.12}$$

And it is the symmetry of the phenomenological coefficients Λ_{kj} which we wish to determine on the basis of (2.2.10).

To carry out our program, it is necessary to convert (2.2.12) into finite difference equations. This conversion requires a new, physical assumption. Suppose that there exists a characteristic time τ which is larger than the time required for the initial accelerations to die out but still much smaller than the time by which the fluctuations have appreciably decayed. It is this time scale which is of interest to us and it is over this time scale that ξ_k and $\dot{\xi}_k$ are assumed to remain 'sensibly' constant. The existence of such a time scale, according to Casimir (1945), is really a new hypothesis.

It is over this time scale that we can write (2.2.12) in a finite difference form. We then average this finite difference equation and obtain

$$\langle \xi_k(t+\tau) - \xi_k(t) \rangle = \tau \Lambda_{kj} \chi_j \tag{2.2.13}$$

Why does not the second member vanish according to (2.2.9)? It is because we are actually performing a non-equilibrium average in (2.2.13), although it has never been stated as such in the literature. The forces can now be eliminated by multiplying (2.2.13) by $\xi_j(t)$ and performing an equilibrium average. In view of (2.2.7) we get

$$\langle \xi_j(t)\{\xi_k(t+\tau) - \xi_k(t)\} \rangle = \tau \Lambda_{kj} k \tag{2.2.14}$$

Switching indices gives

$$\langle \xi_k(t)\{\xi_j(t+\tau) - \xi_j(t)\} \rangle = \tau \Lambda_{jk} k \tag{2.2.15}$$

We now make use of the property that the process is stationary, so that it really makes no difference at what time we begin our observation. In other words, (2.2.10) is invariant with respect to a translation in time, namely

$$\langle \xi_j(t)\xi_k(t+\tau) \rangle = \langle \xi_j(t+\tau)\xi_k(t) \rangle \tag{2.2.16}$$

And consequently, from (2.2.14) and (2.2.15) we obtain

$$\Lambda_{kj} = \Lambda_{jk} \tag{2.2.17}$$

which is Onsager's reciprocal relation.

We have been rather casual with the definitions and assumptions that have been used to derive (2.2.17). We now examine the crucial points of the derivation more carefully. Onsager (1931) states that we would like to think that the dynamical laws governing the world of atoms are reversible in the same sense as the dynamical laws of conservative systems. According to the principle of dynamical reversibility (Tolman, 1938), corresponding to any possible motion of a system, there would be a possible reverse motion such that the same coordinates would be reached in reverse order which is achieved by reversing the velocities. This principle guarantees the existence of reverse states of the motion.

We now ask for the probabilities of finding a molecule in its forward and reverse states. We know that at equilibrium the velocity distribution is Maxwellian so that a mere reversal of the velocities does not alter the energy. From this we conclude that, at equilibrium, any molecular process will occur at the same frequency (or at the same rate) as the reverse of that process, since the probabilities of finding the molecule in the forward and reverse states are the same. This is the gist of the principle of microscopic reversibility at equilibrium (Tolman, 1938). Let us now go back and see what this principle implies about the fluctuations that we have been discussing.

The principle of dynamical reversibility and *not* the principle of microscopic reversibility at equilibrium would require

$$\langle \xi_k(\tau) \rangle = \langle \xi_k\,(-\tau) \rangle \tag{2.2.18}$$

and

$$\langle \dot{\xi}_k(\tau) \rangle = -\langle \dot{\xi}_k(-\tau) \rangle \tag{2.2.19}$$

which is obtained by setting $t = 0$ in (2.2.8). In other words, the average value of the (fluctuating) thermodynamic variable is to be interpreted as a coordinate of an isolated, conservative mechanical system. Condition (2.2.18) is rather demanding; it requires $\langle \xi_k(\tau) \rangle$ and $\langle \xi_k(-\tau) \rangle$ to be solutions of the same equation. This is only possible if the equation is *conservative*. Let L be the Lagrangian function which we assume to be quadratic, or at least even, in the velocities. The conservative equations for forward and reverse motions can then be written in the forms

$$\frac{\mathrm{d}}{\mathrm{d}\tau}\frac{\partial L}{\partial \langle \dot{\xi}_k \rangle} - \frac{\partial L}{\partial \langle \xi_k \rangle} = 0 \qquad \text{for } \langle \xi_k \rangle = \langle \xi_k(\tau) \rangle \tag{2.2.20}$$

and

$$\frac{\mathrm{d}}{\mathrm{d}(-\tau)}\frac{\partial L}{\partial(-\langle \dot{\xi}_k \rangle)} - \frac{\partial L}{\partial \langle \xi_k \rangle} = 0 \qquad \text{for } \langle \xi_k \rangle = \langle \xi_k(-\tau) \rangle \tag{2.2.21}$$

We have already mentioned that equations (2.2.11) are the *average* phenomenological regression laws which are satisfied by $\langle \xi_k(\tau) \rangle$ since the fluctuations are assumed to obey, *on the average*, the empirical laws (Onsager, 1931). These regressions laws are not conservative and therefore are not satisfied by $\langle \xi_k(-\tau) \rangle$. Onsager was well aware of this apparent contradiction.

To make our ideas more concrete let us consider two systems S and S' with coordinates $\langle \xi_k \rangle$ and $\langle \xi_k' \rangle$. The former carries out the forward motion while the latter carries out the reverse motion. If the two systems are conservative then at time $\tau = 0$ we have

$$\langle \xi_k(0) \rangle = \langle \xi_k'(0) \rangle \tag{2.2.22}$$

which shows that their configurations are in agreement while their motions would be in the reverse directions

$$\langle \dot{\xi}_k(0) \rangle = -\langle \dot{\xi}_k'(0) \rangle \tag{2.2.23}$$

Onsager (1931) claims that (2.2.23) vanishes for phenomenological regression laws that are given by (2.2.11)—the velocity has a discontinuity at $\tau = 0$. Hence the phenomenological equations apply only to fluctuations for $\tau > 0$. To support this claim, Onsager argues that this objection is removed when we recognise that (2.2.11) are only approximate descriptions to the process that neglect the time needed for acceleration.

Albeit equations (2.2.11) are only approximations, they still however contain the basic element which causes the accelerations to die out and drive the process to equilibrium. The thermodynamic forces that cause the process to relax to equilibrium will always destroy the symmetry in past and future. It is therefore incomprehensible how the principle of dynamical reversibility (2.2.18) can be applied to the phenomenological regression laws (2.2.11) even for a time $\tau > 0$ which is greater than the time needed for the accelerations to die out.

The basic element which we have overlooked is the stochastic nature of the process. The phenomenological regression laws (2.2.11) can be converted into stochastic differential equations, of the Langevin type, through the addition of random forces. The processes will then turn out to be Markoffian, whose future is independent of its past history but depends only on the present state of the system. Each process will trace out a path in the course of time which, on account of the Markoff property, can be subdivided into an innumerable set of discontinuous path segments, strung end to end. Over any of these intervals, the probability of a forward transition will be equal to the probability of the reverse transition, provided the reference state is that of equilibrium. It is over this small time scale, where the system behaves as a free particle, that we expect the principle of microscopic reversibility to apply. Over a longer time scale, the symmetry in past and future will be destroyed by the attractive nature of the state of equilibrium. This is to say that the irreversibility of the process becomes apparent when our observations become long enough to discern a trend in the evolution of the stochastic process. A detailed exposition would be out of line with the main objectives of this book; let it suffice to say that the proof involves the probability of 'fluctuating paths'.

2.3 Criticisms of classical non-equilibrium thermodynamics

Casimir (1945) pointed out that the class of reciprocal relations (2.2.17) applies specifically to the case in which the thermodynamic fluxes are the time derivatives of the extensive thermodynamic variables. As an illustration, he took the case of thermal conduction which is described by the phenomenological relation

$$\boldsymbol{q} = -\mathbf{K}\boldsymbol{g} \tag{2.3.1}$$

where $\boldsymbol{g}$ is the thermal gradient. Certainly the heat flux $\boldsymbol{q}$ is not the time derivative of a thermodynamic variable, so we cannot conclude that the symmetry of the thermal conductivity tensor $\mathbf{K}$ is a consequence of the Onsager reciprocal relation (2.2.17). As a matter of fact, we could add any vector to (2.3.1) whose divergence is zero, since only the divergence of the heat flux has a physical meaning. The divergence of (2.3.1) is

$$\nabla\cdot\boldsymbol{q} = \hat{\mathbf{K}}:\nabla\boldsymbol{g} + (\nabla\cdot\mathbf{K})\cdot\boldsymbol{g} \tag{2.3.2}$$

where $\hat{\mathbf{K}}$ is the symmetric part of $\mathbf{K}$, namely

$$\hat{\mathbf{K}} = \tfrac{1}{2}(\mathbf{K}+\mathbf{K}^T) \tag{2.3.3}$$

If the material is homogeneous, $\mathbf{K}$ can depend upon the spatial coordinates only through the temperature θ. In this case, (2.3.2) becomes

$$\nabla\cdot\boldsymbol{q} = \hat{\mathbf{K}}:\nabla\boldsymbol{g} + \boldsymbol{g}\cdot\partial_\theta\hat{\mathbf{K}}\boldsymbol{g} \tag{2.3.4}$$

Therefore, if the thermal conductivity tensor is constant or the material is homogeneous, the antisymmetric part of $\hat{\mathbf{K}}$,

$$\tilde{\mathbf{K}} = \tfrac{1}{2}(\mathbf{K}-\mathbf{K}^T) \tag{2.3.5}$$

will be of no physical consequence and it can therefore be put equal to zero. This is little consolation, since the vanishing of (2.3.5) is not a consequence of (2.2.17). Usually, an attempt is made to salvage the Onsager reciprocal relation in these cases by dividing the system up into a small number of cells which is used to express the thermal gradient as a finite difference in temperature and the thermodynamic flux as the time derivative of the temperature (Casimir, 1945; Yourgrau, van der Merwe and Raw, 1966). This is equivalent to the admission that there are no such things as partial differentiation equations in physics (Truesdell, 1969). Moreover, the temperature is not an extensive variable. But even if we were to accept the mathematics, it is still very hard to believe that the symmetry of the thermal conductivity tensor is based on some type of detailed balancing at equilibrium.

In cases where these thermodynamic fluxes are material or heat fluxes, there is no underlying microscopic reason for the phenomenological symmetry. If the symmetry exists then the motion is a 'potential' flow, otherwise not. Consider the quadratic form

$$\boldsymbol{x}\cdot\hat{\mathbf{K}}\boldsymbol{x} = \text{const} \tag{2.3.6}$$

where $\boldsymbol{x}$ is any vector. If $\boldsymbol{x} = \boldsymbol{a}$, a position vector with respect to the origin, then (2.3.6) represents a family of concentric ellipsoids on which the temperature is constant. The derivative of (2.3.6) is parallel to the thermal gradient and the surface of constant (velocity) potential cuts all instantaneous stream lines at right angles. This means that if $\check{\mathbf{K}} = \mathbf{0}$, then the heat flux vector is tangent to the stream lines and there is only irrotational motion. Alternatively, if $\check{\mathbf{K}} \neq \mathbf{0}$, then the motion will be rotational; the flow is no longer potential flow and there does not exist a velocity potential. The expression for the velocity potential is obtained by setting $\boldsymbol{x} = \boldsymbol{g}$ and the heat flux vector is obtained by differentiation with respect to $\boldsymbol{g}$. Therefore, the symmetry of the thermal conductivity tensor is simply an expression of potential flow, nothing more and nothing less.

The red herring of classical non-equilibrium thermodynamics is the question of how the forces and fluxes are to be chosen in the expression for the entropy production. Without some guiding principle, no significance can be attributed to the presence or absence of symmetry of the phenomenological coefficients (Coleman and Truesdell, 1960; Truesdell, 1969). The literature abounds in such types of sterile discussion. One school of thought rejects the question and claims that it is immaterial how the force and fluxes are to be chosen, save that non-singular linear transformations of a particular nature must leave the entropy production invariant. Its major disciple is Meixner (1943). This criterion of admissibility of the thermodynamic forces and fluxes into the bilinear expression of the entropy production has been criticised by Verschaffelt (1951), who produced a linear transformation that left the entropy production invariant but destroyed the symmetry of the matrices of the phenomenological coefficients. In response, Davies (1952) claimed that such symmetry-destroying transformations 'can be of no physical importance'. Hooyman, de Groot and Mazur (1955) have attempted to set things right by invoking the rule that the thermodynamic force (variation) is defined by (2.2.6). This rule is quickly forgotten in de Groot and Mazur (1962) in their attempt to encompass irreversible processes where there is a material flux, e.g. diffusion, heat conduction and viscosity. In these cases, the thermodynamic forces were taken to be the gradient of the conjugate intensive variables.

Another school of thought uses the principle of invariance of the entropy production to claim that any statement regarding the symmetry of the matrix of the phenomenological coefficients is completely vacuous until some rule is decided upon how the thermodynamic forces and fluxes are to be chosen. In other words, there should exist criteria which identify the forces and fluxes other than their entry into a bilinear expression for the entropy production. If such criteria do not exist, then Onsager's reciprocal relations are precarious. The major adherents of this school of thought are Coleman and Truesdell (1960). They show that through a linear combination of the forces and fluxes, an antisymmetric matrix can be added to the phenomenological coefficient matrix which leaves the form of the entropy production invariant. The Coleman and Truesdell result has brought about some degree of concern (Pings and Nebeker, 1965; Hanley, 1969). Let us consider their argument in some detail.

Coleman and Truesdell interpret the reciprocity theorem of Onsager in the following way: consider a set of phenomenological relations between the fluxes and forces

$$\iota_i = \Lambda_{ij}\chi_j \tag{2.3.7}$$

If: (1) the thermodynamic fluxes ι_i are defined as

$$\iota_i \equiv \dot{\xi}_i \tag{2.3.8}$$

(2) (2.2.3) and (3) (2.2.6), *then* (2.2.17) follows as a consequence. Surely Onsager did not say anything to this effect and probably this is why Coleman and Truesdell cite Casimir (1945), de Groot (1952) and Hirschfelder, Curtiss and Bird (1954) instead. But for the sake of argument, let us follow their reasoning to the end.

Expression (2.2.3) is differentiated in time to obtain

$$\dot{\eta} = -S_{ij}\dot{\xi}_i\xi_j \tag{2.3.9}$$

Then comparing this expression with (2.3.8) and (2.2.6), Coleman and Truesdell argue that in order for the fluxes ι_i and forces χ_i in (2.3.7) to obey (2.3.8), (2.2.3) and (2.2.6), it is necessary that

$$\dot{\eta} = -\chi_i\iota_i \tag{2.3.10}$$

Although (2.3.10) follows from (2.3.8), (2.2.3) and (2.2.6), the Coleman – Truesdell claim is that

$$(2.3.7) + (2.3.10) \Rightarrow (2.2.17)$$

is self-contradictory. Their demonstration is based on the fact that the matrix of any quadratic form is always symmetric, which is a trivial result of linear algebra. Consequently, we can construct a linear combination

$$\iota_i' = \iota_i + \Omega_{ij}\chi_j \tag{2.3.11}$$

$$\chi_i' = \chi_i \tag{2.3.12}$$

where

$$\Omega_{ij} = -\Omega_{ji} \tag{2.3.13}$$

which leaves (2.3.9) invariant.

However, what Coleman and Truesdell failed to realise is that (2.2.17) are the exactness conditions for the scalar potential (2.3.9). In other words,

$$(2.3.7)+(2.2.17) \Rightarrow (2.3.10)$$

but *not* vice versa. Had Coleman and Truesdell realised this, it would have saved them the effort of their endeavours. Let us be more specific: the thermodynamic process is governed by (2.3.7) with the understanding (2.3.8). If the motion manifests the symmetry (2.2.17), and it will always do so in linear thermodynamics, then (2.3.10) exists, otherwise not. This is exactly the same as the condition of potential flow which we have previously discussed. Furthermore, the Coleman–Truesdell claim that 'a reciprocity theorem based on (2.3.7), (2.3.8), (2.2.3) and (2.2.6) does not suffer from the logical defects of one based on (2.3.7) and (2.3.10)' is meaningless, since the thermodynamic process alone determines whether conditions (2.2.17) are satisfied. Care should always be taken to distinguish between what is given and what is to be derived.

There is some validity to the remark on how can non-equilibrium thermodynamics treat viscosity, diffusion and heat conduction, since definition (2.3.8) must be strictly adhered to in all cases (Truesdell, 1969). This will be a major concern of ours; at this point let the following remarks suffice: mechanical and thermal phenomena will require different types of analysis, in a way that was already appreciated by Kelvin and Thomson. Mechanical effects of dissipation can be accounted for in an indirect way that expresses the flows and forces in terms of the time derivatives and gradients of a complete set of internal state or 'hidden' variables (cf. section 5.1).

If the major goal of non-equilibrium thermodynamics would be the identification of the χ's and the ι's in the expression for the entropy production, the science would be reduced to a sterile series of algebraic manipulations that are devoid of any physical content. Controversies on how the fluxes and forces are to be selected and the meaning of the symmetry of the phenomenological coefficient matrices, do not produce any constructive or fruitful results. We know what the forces and fluxes are; they are defined by (2.2.5) and (2.3.8), respectively. Then the question is, do Onsager's reciprocal relations (2.2.17) hold in general, and if not, when do they hold? We also know the answer to this question (Lavenda, 1972): they necessarily hold in linear thermodynamics but there is no guarantee that they are valid in nonlinear thermodynamics. In other words, there will always exist scalar non-equilibrium thermodynamic potentials in linear thermodynamics. Whether such scalar potentials exist in nonlinear thermodynamics depends entirely on the thermodynamic process under consideration. As a consequence, certain types of thermodynamic processes are outlawed in linear thermodynamics. This is because linear thermodynamics treats irreversible processes that are occurring in an isolated system. Without energy reservoirs there can be no circulatory or rotational motion.

All these simple, but important, conclusions follow from Onsager's formulation. Non-equilibrium thermodynamics has gone astray in the interpretation and application of the formulation. This is not to say that Onsager has provided us with the complete story or that he was correct in applying his reciprocal relations to problems involving heat conduction. Moreover, we cannot expect that non-equilibrium thermodynamics will ever be a completely autonomous branch of macroscopic physics, since it deals with kinetic processes in which the concept of stability enters. And we know that necessary and sufficient conditions of stability are always given by kinetic stability analysis. We do not want to go into details at this point but rather to display the inadequacies of the classical formulation of non-equilibrium thermodynamics.

One of the three pillars of classical non-equilibrium thermodynamics is Curie's principle (Curie, 1894). The other two pillars are the universal validity of the Onsager reciprocal relations and the existence of a 'local' equilibrium. Let us consider the first of these three, since the remaining two will be taken up in chapters 4, 7 and 8. One interpretation of Curie's principle is 'that macroscopic causes cannot have more elements of symmetry than the effects they produce' (Yourgrau, van der Merwe and Raw, 1966). Another statement of Curie's principle, and probably the original one (Truesdell, 1969), forbids all coupling between quantities of different tensorial character (Prigogine and Mazur, 1951). Other statements of the principle prohibit the 'interaction' and 'coupling' of quantities that are of a different tensorial character (Kirkwood and Crawford, 1952; de Groot, 1952). However, after consulting textbooks on linear algebra, it was realised that the scalar product of two tensors is a scalar, so that if one of the tensors is a phenomenological coefficient matrix, then a tensorial force of order two can definitely interact or couple to a tensor of order zero. This realisation led to a more cautious wording of Curie's principle: 'if the system is isotropic, it may be shown that those terms which correspond to a coupling of tensors whose orders differ by an odd number do not occur' (Hirschfelder, Curtiss and Bird, 1954).

Truesdell (1966) has been the spearhead of the attack on Curie's 'theorem' which he considers to be the 'non-existent theorem of algebra'. Truesdell (1969) offers

$$\boldsymbol{q} = -(\mathbf{K}_0 + \mathbf{K}_1 \theta)\boldsymbol{g} \tag{2.3.14}$$

as a counter-example to Curie's principle. He concludes that 'the scalar θ seems to be interacting or coupling with vectors $\boldsymbol{q}$ and $\boldsymbol{g}$'. Although it could be argued that $\boldsymbol{g}$ and not θ is the force which causes the heat flux $\boldsymbol{q}$, this is not the relevant point in question. Rather, $\boldsymbol{q}$ like any material flux is not uniquely defined and only its divergence has a physical meaning. So whether $\boldsymbol{q}$ 'couples' to any non-conjugate force of a different tensorial order is completely irrelevant.

To illustrate this point, consider a steady state in which there is diffusion and chemical reaction. According to Curie's principle, we would be forced into the conclusion that no coupling between diffusion and chemical reaction can exist in an isotropic system. It is well known in classical non-equilibrium thermodynamics that the set of phenomenological relations are (de Groot and Mazur, 1962).

$$\boldsymbol{J}_i = -\Delta_{ij}\boldsymbol{\nabla}\mu_j \tag{2.3.15}$$

$$\rho\dot{r}_k = -\Gamma_{ki}A_i \tag{2.3.16}$$

where the thermodynamic forces are the gradients of chemical potentials $\boldsymbol{\nabla}\mu_j$ and the chemical affinities A_i which cause the thermodynamic 'fluxes' $\boldsymbol{J}_i$, the diffusion flux, and $\dot{r}_k$, the time rate of change of the extent of reactions, respectively. Since $\boldsymbol{J}_i$ is a vectorial 'flux' whereas $\dot{r}_k$ is a scalar 'flux', Curie's principle forbids a coupling of the chemical reactions to diffusion. This is expressed by the set of phenomenological relations (2.3.15) and (2.3.16) where $\|\Delta_{ij}\|$ and $\|\Gamma_{ki}\|$ are the matrices of the phenomenological coefficients.

Curie's principle has led us astray since by taking the divergence of (2.3.15), multiplying (2.3.16) by the stoichiometric coefficients ν_{ik} and summing over the k reactions, we obtain

$$\nabla \cdot \boldsymbol{J}_i = \rho \sum_k \nu_{ik} \dot{r}_k \tag{2.3.17}$$

which is just the steady state form of the continuity equations. Equations (2.3.17) display quite an active 'interaction' or 'coupling' between diffusion and chemical reactions. From the viewpoint of Curie's principle, it would seem very odd indeed that an interaction could suddenly appear simply by taking the divergence of one of the thermodynamic fluxes. On this basis, we conclude that (2.3.15) is physically meaningless, since it is only the divergence of the material flux which has a physical meaning. Therefore, Curie's principle detracts rather than adds anything new or important to non-equilibrium thermodynamics.

In essence, the entire controversy over Curie's principle is superfluous, since kinetic stability analysis asserts that the seemingly licentious combination of diffusion and chemical reactions is what actually determines the stability of the thermodynamic process. Furthermore, any result of non-equilibrium thermodynamics must be compatible with that obtained from kinetic stability analysis.

This concludes the survey of classical non-equilibrium thermodynamics. Chapter 3 proceeds to the more recent theories of non-equilibrium thermodynamics that claim a far greater generality than what has hitherto been achieved by classical non-equilibrium thermodynamics.

References

Casimir, H. B. G. (1945). On Onsager's principle of microscopic reversibility. *Rev. mod. Phys.*, **17**, 343–350.

Coleman, B. D. and Truesdell, C. (1960). On the reciprocal relations of Onsager. *J. chem. Phys.*, **33**, 28–31.

Curie, P. (1894). Sur la symétrie dans les phénomènes physique, symétrie d'un champ électrique et d'un champ magnétiques. *J. Phys. Radium*, **3**, 393–415.

Davies, R. O. (1952). Transformation properties of the Onsager relations. *Physica*, **18**, 182.

de Groot, S. R. (1952). *Thermodynamics of Irreversible Processes*, North-Holland, Amsterdam.

de Groot, S. R. and Mazur, P. (1962). *Non-equilibrium Thermodynamics*, North-Holland, Amsterdam.

Einstein, A. (1910). *Annln Phys.*, **33**, 1275.

Fowler, R. H. (1929). *Statistical Mechanics*, Cambridge University Press, Cambridge.

Hanley, H. J. M. (1969). *Transport Phenomena in Fluids*, ch. 3 (Ed. Hanley, H. J. M.), Marcel Dekker, New York.

Hirschfelder, J. O., Curtiss, C. F. and Bird, R. B. (1954). *Molecular Theory of Gases and Liquids,* Wiley, New York.

Hooyman, G. J., de Groot, S. R. and Mazur, P. (1955). *Physica,* **21,** 360.

Katchalsky, A. and Curran, P. F. (1965). *Nonequilibrium Thermodynamics in Biophysics,* Harvard University Press, Cambridge, Mass.

Kelvin, Lord (W. Thomson) (1854). *Proc. R. Soc. Edinb.* p. 123.

Kirkwood, J. G. and Crawford, Jr, B. (1952). The macroscopic equations of transport. *J. phys. Chem.,* **56,** 1048–1051.

Landau, L. D. and Lifshitz, E. M. (1958). *Statistical Physics,* Pergamon Press, Oxford.

Lavenda, B. H. (1972). Concepts of stability and symmetry in irreversible thermodynamics I. *Found. Phys.,* **2,** 161–179.

Meixner, J. (1943). Zur Thermodynamik der irreversiblen Prozessen in Gasen mit chemisch reagierenden, dissozierenden und anregbaren Komponenten. *Annln Phys.,* **43,** 224–270.

Onsager, L. (1931). Reciprocal relations in irreversible processes I. *Phys. Rev.* **37,** 405–426; Reciprocal relations in irreversible processes II. *Phys. Rev.,* **38,** 2265–2279.

Pings, C. J. and Nebeker, E. B. (1965). *I and EC Fundamentals,* **4,** 376.

Prigogine, I. (1967). *Thermodynamics of Irreversible Processes,* 3 ed., Wiley-Interscience, New York.

Prigogine, I. and Mazur, P. (1951). Sur deux formulations de l'hydrodynamique et le problème de l'hélium liquide II. *Physica,* **17,** 661–679.

Rayleigh, Lord (J. W. Strutt) (1873). *Proc. Lond. math. Soc.,* **4,** 357–363.

Rayleigh, Lord (J. W. Strutt) (1877). *Theory of Sound,* Macmillan, London.

Thomson J. J. (1888). *Applications of Dynamics to Physics and Chemistry,* Macmillan, London.

Tolman, R. C. (1938). *The Principles of Statistical Mechanics,* Oxford University Press, London.

Truesdell, C. (1966). *Six Lectures on Modern Natural Philosophy*, Springer-Verlag, New York.

Truesdell, C. (1969). *Rational Thermodynamics,* McGraw-Hill, New York.

Verschaffelt, J. E. (1951). Sur l'effet thermomécanique. *Bull. Acad. Roy. Belg. Cl. Sci.,* **37,** 853–872.

Wei, J. (1966). Irreversible thermodynamics in engineering. *Ind. Engng Chem.,* **58,** 55–60.

Yourgrau, W., van der Merwe, A., and Raw, G. (1966). *Treatise on Irreversible and Statistical Physics,* Macmillan, New York.

3 'Rational' Thermodynamics

One school of non-equilibrium thermodynamics has been created by the inclusion of thermodynamic-like concepts in the mechanics of the continua. It has been named, by Truesdell (1966, 1969), the school of 'rational' thermodynamics. Rational thermodynamics deals essentially with two types of materials: elastic materials with linear viscosity and elastic materials with 'memory'. The basic tenets of the thermodynamics of viscoelastic materials has been formulated by Coleman and Mizel (1964), and the thermodynamic analysis of elastic materials in which the present state of the stress is influenced by the history of the strain, has been given by Coleman (1964*a*, 1964*b*).

The two theories share a common outlook: the employment of a formal statement of the second law as a restriction on the form of the 'constitutive' equations. They differ in regard to their handling and interpretation of dissipation. A third way of accounting for dissipative effects has been proposed by Coleman and Gurtin (1967). In Coleman and Mizel's analysis, dissipation is the result of viscous forces.In the Coleman theory, dissipation arises from the history of the deformed material. And finally, in Coleman and Gurtin's approach, dissipation is accounted for in the internal state variables. Although it has been claimed (Truesdell, 1969) that these three approaches are separate and distinct, applying to different classes of materials, we shall see that all three theories provide alternative expressions for the dissipation function. This is what we would expect logically: independent of the particular characteristics of the material, dissipation is always to be implicated with the rates of strain.

The fundamental idea underlying all the work in rational thermodynamics is that a formal statement of the second law serves as a restriction on the constitutive relations. The idea can be traced back to Eckart (1940), who realised the importance of treating the second law as a restriction on the form of phenomenological equations as opposed to the derivation of the second law from laws which are less general. Whereas Eckart made a clear distinction between the level of importance of the Gibbs equation in contrast to the phenomenological laws, such as Fourier's law of heat conduction, no such distinction is made in rational thermodynamics. Motivation for this approach is to be found in a statement made by Coleman and Mizel (1964). Citing the analyses of Prigogine (1949) of the Boltzmann equation, using the

Chapman – Enskog procedure, as a microscopic justification of first order thermodynamics, these authors claimed that if non-equilibrium thermodynamics was to be an autonomous branch of continuum physics, then justification for the use of the Gibbs relation, or for that matter any thermodynamic relation, must ultimately come from the macroscopic theory itself.

Since the Gibbs equation is not assumed to be valid at the outset, the thermodynamic potentials are reduced to the level where they are to satisfy constitutive relations similar to those of the stress tensor and heat flux vector. Moreover, in the absence of the Gibbs equation, no distinction can be made between equations of state and the phenomenological laws that relate the motion to the 'causes' or the constitutive dependent variables. Everything is lumped together under the all-encompassing heading of 'constitutive' equations. The ultimate form of the constitutive equations is to be determined from a formal statement of the second law, the so-called 'Clausius – Duhem' inequality, and their compatibility with the macroscopic balance equations of energy and momentum.

In this chapter, the theory of rational thermodynamics will be critically analysed. We must always be careful to bear in mind the philosophy of rational thermodynamics: the Clausius – Duhem inequality is assumed always to hold. It is not to be interpreted as a criterion of stability, although it is used to derive stability criteria in terms of the well-known extrema of thermodynamic potentials (Erickson, 1966). This we know to be true only in the immediate vicinity of equilibrium and we thus obtain a clue as to the limitations of the theory. Consequently, the Clausius – Duhem inequality is to be treated on equal footing with the fundamental laws of physics.

Such a vast literature has grown up around rational thermodynamics that we cannot hope to treat every detail of the theory. Instead we will seek out the fundamental and crucial points of the theory that will ultimately determine the validity of rational thermodynamics. For the interested reader, a full list of references is to be found in Truesdell's latest monograph (Truesdell, 1969). A more concise presentation of Coleman's analysis of materials with memory (Coleman, 1964*a*, 1964*b*) has been given by Gurtin (1968). A more recent review of the state of affairs of the thermodynamics of continuous media is to be found in the work of Eringen (1975). These references suffice to give a general, overall picture of the developments in rational thermodynamics.

3.1 Axioms of rational thermodynamics

It is fashionable, in rational thermodynamics, to expound the theory in a number of axioms. We shall briefly consider some of the more crucial axioms that will concern us in the following paragraphs.

3.1.1 Axiom of 'admissibility'

An admissible thermodynamic process is one in which the constitutive relations satisfy the Clausius – Duhem inequality and are consistent with the macroscopic balance equations of energy and momentum.

The Clausius – Duhem inequality is considered in rational thermodynamics to be a formal statement of the second law. It is derived simply by eliminating the divergence of the heat flux vector between the balance equations for the entropy and internal energy. However, there exists a peculiarity in the formation of these two balance equations which deserves some comment. Truesdell and Noll (1965) have expressed the second law in its global form:

$$\frac{d}{dt}\int_B \rho\eta \, dv + \int_{\partial B} \boldsymbol{q}\cdot\boldsymbol{n} \, dS - \int_B \rho\frac{r}{\theta} dv = \sum_S \tag{3.1.1}$$

The doubtful point concerning their formulation of the second law lies in the specification of a 'radiation' energy per unit mass and time, r, that is absorbed by the body B. If it is a heat supply then it has to pass over the surface ∂B and so it should be included in the heat flux term. On the contrary, if it is a volume source term, then it should be accounted for in the entropy source term. However, as we shall soon see, the radiation energy is required for the self-consistency of the theory—it does not determine the thermodynamic process but, rather, the thermodynamic process determines it! Still more perplexing is the fact that it vanishes in the final statement of the second law that is used in rational thermodynamics.

Under the appropriate continuity conditions, we can write

$$\sum_S = \int_B \rho\sigma_S \, dv$$

Then, in order for (3.1.1) to be valid for all parts of the body B, it is necessary and sufficient that

$$\dot{\eta} + \frac{1}{\rho}\nabla\cdot(\boldsymbol{q}/\theta) - \frac{r}{\theta} = \sigma_S \geq 0 \tag{3.1.2}$$

Needless to say, we are dealing here with a particular form of the entropy balance equation that is restricted to 'simple' materials for which the entropy flux is just $(\boldsymbol{q}/\theta)$. For a generalisation of the entropy flux expression see Müller (1967).

The local form of the entropy balance equation (3.1.2) is supplemented by the balance equations of energy and momentum

$$\dot{\varepsilon} + \frac{1}{\rho}\nabla\cdot\boldsymbol{q} - r = \frac{1}{\rho}tr\{\mathbf{TL}\} \tag{3.1.3}$$

$$\ddot{x} - \frac{1}{\rho}\nabla\cdot\boldsymbol{T} = b \tag{3.1.4}$$

where again we note the presence of the ubiquitous radiation energy term in the internal energy balance equation (3.1.3). If we now eliminate the divergence of the heat flux between equations (3.1.2) and (3.1.3), we come out with

$$\sigma_S = \dot{\eta} - \dot{\varepsilon}/\theta + \frac{1}{\rho\theta}tr\{\mathbf{TL}\} - \frac{\boldsymbol{q}\cdot\boldsymbol{g}}{\rho\theta^2} \geq 0 \tag{3.1.5}$$

which is commonly referred to as the Clausius–Duhem inequality. In one sense we should be happy that the radiation energy has been eliminated in the process of deriving (3.1.5); otherwise, it could have violated the inequality. Another word of caution concerning the Clausius–Duhem inequality (3.1.5) is that it is valid for 'simple' materials only, i.e. those in which the entropy flux is due solely to heat conduction (cf. section 9.1).

3.1.2 Axiom of determinism

All constitutively dependent variables are functionals of the history of the motion and the temperature. This axiom generalises the mechanics of the continua to include thermodynamic considerations. In particular, we would like to know the effect of the temperature on the stress. It is especially pertinent to Coleman's analysis of elastic materials with memory (Coleman, 1964*a*, 1964*b*).

3.1.3 Axiom of equipresence

This axiom is commonly accredited to Truesdell (Truesdell and Toupin, 1960) and it is basically an anti-discriminatory postulate. It states in essence that, at the outset, we cannot discriminate among the variables that appear in the individual constitutive equations. If a variable appears in one constitutive relation then there is no *a priori* reason why it should not appear in all the other constitutive relations. The condition for the presence or absence of a constitutive independent variable is to be determined by the Clausius – Duhem inequality (3.1.5). The compatibility of the resulting constitutive relation with the macroscopic balance equations of energy and momentum can always be ensured, since we have two parameters at our disposal, r and b.

The axiom of equipresence destroys all thermodynamic functional dependencies and representations. Moreover, it necessitates the construction of mechanical machinery for weeding out unwanted functional dependencies. There is a wealth of literature on the subject. However, it always works out that the Clausius – Duhem inequality selects the free energy as the thermodynamic potential and the entropy as the derived thermodynamic variable. It is then no surprise that Truesdell's axiom on equipresence works: the derived thermodynamic variable will always have the same variables as the thermodynamic potential, since its constitutive relation is just a generalised equation of state. Where the axiom of equipresence seems to be of some value is in the determination of the constitutive relations for the non-thermodynamic variables, such as the stress tensor and the heat flux vector.

Caution must be exercised in applying the formalism of rational thermodynamics: it can only be applied to *independent* thermodynamic processes. It fails when (1) a coupling exists between thermodynamic processes, such as a viscoelastic solid that conducts heat, or (2) the constitutive independent variables and their time derivatives cannot be varied independently. However, there exists another axiom that guarantees their independence. It has been shown by Edelen and McLennan (1973) that this axiom is a 'convenience' rather than a principle.

3.1.4 Axiom of material frame-indifference or objectivity

This axiom states that the constitutive equations must be invariant under the non-finite Euclidean group. The Euclidean group consists of all transformations of the present configuration $\boldsymbol{x}(X, t)$

$$\boldsymbol{x}'(X, t') = \boldsymbol{c}(t) + \mathbf{Q}(t)\boldsymbol{x}(X, t), \qquad t' = t + a \tag{3.1.6}$$

where a is a constant, $\boldsymbol{c}(t)$ is an arbitrary vector valued function of time and $\mathbf{Q}(t)$ is an arbitrary tensor valued function of time. Since $\boldsymbol{c}(t)$ and $\mathbf{Q}(t)$ carry arbitrary time dependencies, the constitutive independent variables and their time derivatives can be made independent of one another. However, as Edelen and McLennan (1973) demonstrate, the axiom of material frame-indifference eliminates real, physical processes that are invariant under Galilei transformations. Moreover, the axiom of material frame-indifference does not follow from Galilei invariance—it has a content all of its own.

Additional axioms of rational thermodynamics concerning notions of 'memory' will be discussed in section 3.3, since there is no universally accepted definition of memory. Memory means different things to different people and using Eringen's definition of 'smooth' memory (Eringen, 1965), we will be able to show that Coleman and Mizel's analysis of viscoelastic materials and Coleman's analysis of materials with memory are not separate theories as Coleman has claimed (Coleman, 1964*a*). The dilemma is centred on what is meant by dissipation (section 3.3), but before discussing this question it is necessary to understand the formalism.

3.2 The formalism of rational thermodynamics

There is a single, basic formalism of rational thermodynamics that is used to determine the functional forms of the constitutive relations. The net results are: the temperature gradient and the rate dependent variables all drop out of the constitutive relation for the Helmholtz free energy; the free energy determines only the elastic part of the stress. These results, suprisingly enough, are all consistent with the postulate concerning the validity of the generalised Gibbs equation for non-equilibrium processes. However, these views are diametrically opposed to one another. Rational thermodynamics liberates the form of the constitutive relation for the thermodynamic potential and demands that the second law serve as the criterion. 'Generalised' thermodynamics, which will be discussed in the following chapter, restricts the constitutive relation by postulating the validity of the Gibbs equation. Consequently the second law is now free to act as a stability criterion for the thermodynamic process. We shall now consider how rational thermodynamics functions in practice.

At the outset, we assume that a viscoelastic material can be described by the following constitutive equations

$$\varepsilon = \varepsilon(\theta, \ \boldsymbol{g}, \ \mathbf{F}, \ \dot{\mathbf{F}}) \tag{3.2.1}$$

$$\eta = \eta(\theta, \ \boldsymbol{g}, \ \mathbf{F}, \ \dot{\mathbf{F}}) \tag{3.2.2}$$

$$\mathbf{T} = \mathbf{T}(\theta, \ \boldsymbol{g}, \ \mathbf{F}, \ \dot{\mathbf{F}}) \tag{3.2.3}$$

$$\boldsymbol{q} = \boldsymbol{q}(\theta, \ \boldsymbol{g}, \ \mathbf{F}, \ \dot{\mathbf{F}}) \tag{3.2.4}$$

For materially homogeneous materials, such as fluids, it is not necessary to distinguish the function from its value, since the function is independent of a specific reference configuration. When it is necessary to distinguish the function from its value, for example in a materially inhomogeneous solid, we will attach the superscript (^) above the function.

The constitutive equations (3.2.1–4) are by no means the most general ones we could invent. A more general form would be (Eringen, 1965).

$$J[\mathbf{T}, \eta, \varepsilon, \boldsymbol{q}] = 0 \tag{3.2.5}$$

which may not be solvable for a particular constitutive dependent variable or may give rise to a multiplicity of solutions. Hence, the constitutive equations (3.2.1–4) are a particularly simple set of relations that cannot be used in the description of hysteretic, viscoplastic materials, etc.

Moreover, there is nothing unique about the set of constitutive relations (3.2.1–4). Assuming that

$$\frac{\partial}{\partial \theta} \varepsilon(\theta, \boldsymbol{g}, \mathbf{F}, \dot{\mathbf{F}}) \neq 0 \tag{3.2.6}$$

we can solve for θ as a function of ε and write

$$\theta = \theta(\varepsilon, \boldsymbol{g}, \mathbf{F}, \dot{\mathbf{F}}) \tag{3.2.7}$$

On account of Truesdell's axiom of equipresence, the other constitutive relations in this 'representation' are

$$\eta = \eta(\varepsilon, \boldsymbol{g}, \mathbf{F}, \dot{\mathbf{F}}) \tag{3.2.8}$$

$$\mathbf{T} = \mathbf{T}(\varepsilon, \boldsymbol{g}, \mathbf{F}, \dot{\mathbf{F}}) \tag{3.2.9}$$

$$\boldsymbol{q} = \boldsymbol{q}(\varepsilon, \boldsymbol{g}, \mathbf{F}, \dot{\mathbf{F}}) \tag{3.2.10}$$

Observe that whereas the constitutive relations (3.2.1–4) would correspond to the classic equations of state, constitutive relations (3.2.7–10) are those that would appear in the entropy representation (cf. section 1.2). When the thermodynamic, constitutive dependent variables are chosen properly, it will always turn out that the pair of thermodynamic functions will consist of the thermodynamic potential and the thermodynamic derivative of the potential. For example, in the set of constitutive relations (3.2.7–10), the thermodynamic potential is η and its derivative with respect to ε defines the temperature

$$\frac{\partial}{\partial \varepsilon} \eta(\varepsilon, \boldsymbol{g}, \mathbf{F}, \dot{\mathbf{F}}) = [\theta(\varepsilon, \boldsymbol{g}, \mathbf{F}, \dot{\mathbf{F}})]^{-1} \tag{3.2.11}$$

It is not surprising that Truesdell's axiom of equipresence works here: the constitutive relation for θ is just a generalised equation of state.

In the first set of constitutive relations (3.2.1–4), the thermodynamic, constitutive dependent variables have not been chosen correctly. This is apparent when we try to substitute the total derivatives of η and ε into the Clausius–Duhem inequality (3.1.5). When this is done, we realise immediately that the 'natural' thermodynamic potential is the specific Helmholtz free energy

$$\psi = \varepsilon - \eta\theta \tag{3.2.12}$$

Albeit, we could have saved some effort by using the thermodynamic result that the thermodynamic potential of the selected, constitutive independent variables is the Legendre transform of the internal energy with respect to the entropy.

Inserting the specific Helmholtz free energy into the Clausius–Duhem inequality (3.1.5) results in

$$\dot{\psi} + \eta\dot{\theta} + \frac{1}{\rho\theta}\boldsymbol{q}\cdot\boldsymbol{g} - \frac{1}{\rho}tr\{\mathbf{TL}\} \leq 0 \tag{3.2.13}$$

Hence, the constitutive relation

$$\psi = \psi(\theta, \boldsymbol{g}, \mathbf{F}, \dot{\mathbf{F}}) \tag{3.2.14}$$

replaces the constitutive relation for the specific internal energy (3.2.1). Again note that ψ is the thermodynamic potential and its derivative with respect to θ is

$$\frac{\partial}{\partial\theta}\psi(\theta, \boldsymbol{g}, \mathbf{F}, \dot{\mathbf{F}}) = -\eta(\theta, \boldsymbol{g}, \mathbf{F}, \dot{\mathbf{F}}) \tag{3.2.15}$$

which is used as the definition of the conjugate, thermodynamic variable.

The assumption that ψ is a smooth function of the constitutive independent variables ensures the existence of its total derivative

$$\dot{\psi} = \partial_\theta\,\psi\dot{\theta} + \partial_{\boldsymbol{g}}\,\psi\dot{\boldsymbol{g}} + \partial_{\mathbf{F}}\psi : \dot{\mathbf{F}} + \partial_{\dot{\mathbf{F}}}\,\psi : \ddot{\mathbf{F}} \tag{3.2.16}$$

where $\partial_\theta\,\psi$ is a scalar, $\partial_{\boldsymbol{g}}\,\psi$ is the covariant vector of the gradient of ψ with respect to $\boldsymbol{g}$, while $\partial_{\mathbf{F}}\psi$ and $\partial_{\dot{\mathbf{F}}}\psi$ are tensors that are the gradients of ψ with respect to $\mathbf{F}$ and $\dot{\mathbf{F}}$, respectively. Introducing (3.2.16) into the Clausius–Duhem inequality (3.2.13) yields

$$(\partial_\theta\,\psi + \eta)\dot{\theta} + \frac{1}{\rho\theta}\boldsymbol{q}\cdot\boldsymbol{g} + (\partial_{\mathbf{F}}\psi - \mathbf{S}) : \dot{\mathbf{F}} + \partial_{\dot{\mathbf{F}}}\psi : \ddot{\mathbf{F}} + \partial_{\boldsymbol{g}}\,\psi\cdot\dot{\boldsymbol{g}} \leq 0 \tag{3.2.17}$$

where the Piola–Kirchhoff stress tensor is defined (Truesdell, 1966)

$$\mathbf{S} = \rho^{-1}\mathbf{T}(\mathbf{F}^T)^{-1} \tag{3.2.18}$$

This follows from the relation between the rate of strain tensor $\mathbf{L}$ and $\dot{\mathbf{F}}$, $\dot{\mathbf{F}} = \mathbf{LF}$ or

$$\mathbf{L} = \dot{\mathbf{F}}\mathbf{F}^{-1} \tag{3.2.19}$$

provided, of course, det $\mathbf{F} \neq 0$.

We now demand that the Clausius–Duhem inequality (3.2.17) act as a restriction for the constitutive relations (3.2.2–4) and (3.2.14). There are seven variables in the inequality (3.2.17) that can vary: the constitutive independent variables θ, $\boldsymbol{g}$, $\mathbf{F}$ and $\dot{\mathbf{F}}$, plus their time derivatives $\dot{\theta}$, $\dot{\boldsymbol{g}}$, $\ddot{\mathbf{F}}$. In order for the theory to work, these seven quantities must vary independently. The argument employed is based on the principle of material frame-indifference or objectivity (axiom 3.1.4). This means that the constitutive equations must be invariant under Euclidean transformations. As we have already remarked, this does not follow from the known Galilei invariance of the laws of motion. Moreover, Edelen and McLennan (1973) have shown that the constitutive relations for the stress and energy flux, derived from the Boltzmann equation, violate the principle of material frame-indifference while exhibiting invariance under Galilei transformations. Consequently, these authors conclude that the principle of material frame-indifference is to be regarded rather as a 'convenience' than a principle.

Notwithstanding this criticism, the principle of material frame-indifference is used throughout rational thermodynamics. The usual argument runs somewhat as follows: we know the temperature at the material point X' at time t_0 and want to calculate (X, t). We choose the time t_0, a scalar a and a vector $\boldsymbol{a}$, arbitrarily. The time dependent temperature distribution is defined by

$$\theta\,(X, t) = \theta_0 + (t - t_0)a + [(\boldsymbol{g}_0 + (t - t_0)\boldsymbol{a}] \cdot \mathbf{F}_0 \cdot [X - X'] \tag{3.2.20}$$

for $t \geq t_0$ but t sufficiently close to t_0 so that (3.2.20) will be a valid definition of the time dependent temperature distribution. We also have that

$$\theta(X', t_0) = \theta_0, \quad \theta(X', t_0) = a \tag{3.2.21}$$

and

$$\boldsymbol{g}(X', t_0) = \boldsymbol{g}_0, \quad \boldsymbol{g}(X', t_0) = \boldsymbol{a} \tag{3.2.22}$$

which shows that θ and $\dot{\theta}$ and $\boldsymbol{g}$ and $\dot{\boldsymbol{g}}$ can be assigned independently. Analogous conclusions can be deduced concerning the independency of $\mathbf{F}$, $\dot{\mathbf{F}}$ and $\ddot{\mathbf{F}}$.

Accepting the principle of material frame-indifference, we can then assign arbitrary values to the set of constitutive independent variables and their time derivatives. Then $\psi, \eta, \varepsilon(= \psi + \theta\eta)$, $\mathbf{T}$ and $\boldsymbol{q}$ can be computed from the constitutive relations (3.2.14) and (3.2.2–4). However, we must not forget to take into consideration the balance relations (3.1.3) and (3.1.2) which relate ε, $\mathbf{T}$ and $\boldsymbol{q}$ and η, $\boldsymbol{q}$ and θ for the arbitrary assigned values of θ, $\boldsymbol{g}$, $\dot{\mathbf{F}}$, $\dot{\theta}$, $\dot{\boldsymbol{g}}$, $\ddot{\mathbf{F}}$. The condition that the constitutive dependent variables be compatible with the balance equations for arbitrary values of the constitutive independent variables and their time derivatives would be too stringent of a condition if it were not for the unknown quantity r, the specific radiation energy. Coleman and Mizel (1964) contend that once the constitutive dependent variables have been computed, the specific radiation energy can then be determined by the balance laws. Consequently, although the physical significance of the radiation energy—that is, neither an energy flux across the bounding surface nor a heat source—is not clear, the mathematical expedience for the introduction of a radiation energy term into the balance equations is plain enough: it permits the computed constitutive dependent variables to be compatible with the macroscopic balance equations.

The procedure for the elimination of the incorrect functional dependencies in the constitutive relations (3.2.2–4) and (3.2.14) now becomes routine. According to Fourier's law, heat can never flow in the direction of its temperature gradient. The term $\boldsymbol{q} \cdot \boldsymbol{g}$ will never be positive and we can split inequality (3.2.17) into

$$\partial_{\boldsymbol{g}}\psi \cdot \dot{\boldsymbol{g}} + C_1(\theta, \dot{\theta}, \boldsymbol{g}, \mathbf{F}, \dot{\mathbf{F}}, \ddot{\mathbf{F}}) \leq 0 \tag{3.2.23}$$

On the basis of the principle of material frame-indifference, we can arbitrarily assign θ, $\dot{\theta}$, $\boldsymbol{g}$, $\mathbf{F}$, $\dot{\mathbf{F}}$ and $\ddot{\mathbf{F}}$ which gives $C_1(\theta, \dot{\theta}, \boldsymbol{g}, \mathbf{F}, \dot{\mathbf{F}}, \ddot{\mathbf{F}})$ some fixed finite value, say c_1. We thus require

$$\partial_{\boldsymbol{g}}\psi \cdot \dot{\boldsymbol{g}} + c_1 \leq 0 \tag{3.2.24}$$

which is linear in $\dot{\boldsymbol{g}}$. The only way to satisfy the inequality for all values of c_1 is to require

$$\partial_{\boldsymbol{g}}\psi = \boldsymbol{0} \tag{3.2.25}$$

where $\boldsymbol{0}$ is the zero vector. The $\boldsymbol{g}$ dependency drops out of the constitutive relation (3.2.14) and we are left with

$$\psi = \psi(\theta, \mathbf{F}, \dot{\mathbf{F}}, \ddot{\mathbf{F}}) \tag{3.2.26}$$

Therefore, without any loss of generality we can consider henceforth an isothermal process, $\boldsymbol{g} = \boldsymbol{0}$.

It is now convenient to split the reduced form of inequality (3.2.17)

$$(\partial_\theta \psi + \eta)\dot{\theta} + (\partial_{\mathbf{F}}\psi - \mathbf{S}) : \dot{\mathbf{F}} + \partial_{\dot{\mathbf{F}}}\psi : \ddot{\mathbf{F}} \leq 0 \tag{3.2.27}$$

into

$$\partial_{\dot{\mathbf{F}}}\psi : \ddot{\mathbf{F}} + C_2(\theta, \dot{\theta}, \mathbf{F}, \dot{\mathbf{F}}) \leq 0 \tag{3.2.28}$$

Using the same line of reasoning as before, we assign arbitrarily θ, $\dot{\theta}$, $\mathbf{F}$ and $\dot{\mathbf{F}}$ so that (3.2.28) is linear in $\ddot{\mathbf{F}}$. Then the only way that the inequality can be satisfied indentically is to require

$$\partial_{\dot{\mathbf{F}}}\psi = \mathbf{0} \tag{3.2.29}$$

where $\mathbf{0}$ is the zero tensor. In view of conditions (3.2.25) and (3.2.29), the constitutive relation (3.2.14) reduces to

$$\psi = \psi(\theta, \mathbf{F}) \tag{3.2.30}$$

With the free energy given by the constitutive relation (3.2.30), the remaining terms in inequality (3.2.27)

$$(\partial_\theta \psi - \eta)\dot{\theta} + (\partial_{\mathbf{F}}\psi - \mathbf{S}) : \dot{\mathbf{F}} \leq 0 \tag{3.2.31}$$

are linear in $\dot{\theta}$ and $\dot{\mathbf{F}}$. Since these quantities are arbitrary, the only way to satisfy the inequality identically is to set the coefficient of these terms equal to zero:

$$\partial_\theta \psi(\theta, \mathbf{F}) = -\eta(\theta, \mathbf{F}) \tag{3.2.32}$$

$$\partial_{\mathbf{F}}\psi(\theta, \mathbf{F}) = \mathbf{S}(\theta, \mathbf{F})\big|_{\substack{\dot{\mathbf{F}} = \mathbf{0} \\ \boldsymbol{g} = \boldsymbol{0}}} \tag{3.2.33}$$

Therefore what we have succeeded in showing is that the specific free energy determines the specific entropy and the rate independent part of the stress tensor. Furthermore, on account of the conditions (3.2.25), (3.2.29), (3.2.32) and (3.2.33), the Clausius–Duhem inequality (3.2.17) reduces to the commonly referred to 'heat conduction' inequality

$$\boldsymbol{q} \cdot \boldsymbol{g} \leq 0 \tag{3.2.34}$$

3.3 The interpretations of 'dissipation'

In one way, we have generalised equilibrium thermodynamics by showing that the free energy determines the rate independent part of the stress. Alternatively, we could have added the generalised work $\mathbf{S} : \mathbf{F}$ to the Gibbs equation. In either case the functional dependency of the stress tensor does not permit it to depend upon rate dependent variables. As yet the thermodynamic theory cannot give an account of viscoelastic materials. An empirical expression, of rather general validity, relates the stress tensor $\mathbf{T}$ of a homogeneous, isotropic fluid to the rates of strain by the Newtonian law

$$\mathbf{T} = -p\mathbf{I} + 2\nu\mathbf{D} + \lambda(tr\,\mathbf{D})\mathbf{I} \tag{3.3.1}$$

Hence we are confronted with the problem of whether the thermodynamic theory can be generalised in a way that would make it applicable to linear viscoelastic materials.

The viscoelastic thermodynamic theory of Coleman and Mizel (1964) takes the point of view that the thermodynamic potentials can be defined as functions of the state variables which do not include the rate variables and do not depend on the history of the deformed material. Essentially, this is the Gibbsian point of view in which the thermodynamic potentials can be determined solely from a set of static measurements (cf. chapter 5, introduction).

The stress appearing in the energy balance equation (3.1.3) is completely general. In view of (3.2.33), the free energy can determine only the elastic part of the stress. It will therefore prove convenient to decompose the stress $\mathbf{T}$ into an elastic part $\mathbf{T}_E$ and a dissipative part $\mathbf{T}_D$

$$\mathbf{T}(\theta, \boldsymbol{g}, \mathbf{F}, \dot{\mathbf{F}}) = \mathbf{T}_E(\theta, \mathbf{F}) + \mathbf{T}_D(\theta, \boldsymbol{g}, \mathbf{F}, \dot{\mathbf{F}}) \tag{3.3.2}$$

where

$$\rho \mathbf{F}\, \partial_{\mathbf{F}} \psi(\theta, \mathbf{F}) = \mathbf{T}_E(\theta, \mathbf{F})$$

The Clausius – Duhem inequality (3.2.13) can now be written more explicitly as

$$\dot{\psi} + \eta\dot{\theta} + \frac{1}{\rho\theta}\boldsymbol{q}\cdot\boldsymbol{g} - \frac{1}{\rho} tr\{[\mathbf{T}_E(\theta, \mathbf{F}) + \mathbf{T}_D(\theta, \boldsymbol{g}, \mathbf{F}, \dot{\mathbf{F}})]\mathbf{D}\} \leq 0 \tag{3.3.3}$$

where we have used the fact that the Cauchy stress tensor is symmetric: $tr\{\mathbf{TL}\} = tr\{\mathbf{TD}\}$. The effect of introducing the total derivative

$$\dot{\psi} = -\eta\dot{\theta} + \frac{1}{\rho} tr\{\mathbf{T}_E\, \mathbf{D}\} \tag{3.3.4}$$

is to eliminate the non-dissipative terms in inequality (3.3.3)

$$\frac{1}{\theta}\boldsymbol{q}\cdot\boldsymbol{g} - tr\{\mathbf{T}_D\, D\} \leq 0 \tag{3.3.5}$$

In general, $\mathbf{T}_D$ will be a function of $\boldsymbol{g}$ and the two terms in inequality (3.3.5) are coupled through $\boldsymbol{g}$. In the limiting case $\boldsymbol{g} = \boldsymbol{0}$, we obtain the so-called 'mechanical dissipation inequality

$$tr\{\mathbf{T}_D(\theta, \boldsymbol{0}, \mathbf{F}, \dot{\mathbf{F}})\mathbf{D}\} \geq 0 \tag{3.3.6}$$

and in the other limiting case, $\dot{\mathbf{F}} = \mathbf{0}$, we get the heat conduction inequality (3.2.34). The heat flux vector will still be a function of $\theta, \boldsymbol{g}$ and $\mathbf{F}$, so that we will have a generalisation of Fourier's heat conduction law in which the thermal conductivity is a function of θ and $\mathbf{F}$

$$\boldsymbol{q} = -\kappa(\theta, \mathbf{F})\boldsymbol{g} \tag{3.3.7}$$

Inserting (3.3.7) into the heat conduction inequality (3.2.34) yields the quadratic form

$$-\kappa(\theta, \mathbf{F})\boldsymbol{g}^2 \leq 0 \tag{3.3.8}$$

which requires $\kappa(\theta, \mathbf{F}) > 0$.

For the phenomenological relation (3.3.1), the dissipative part of the stress is

$$\mathbf{T}_D = 2\nu\mathbf{D} + \lambda(tr\mathbf{D})\mathbf{I} \tag{3.3.9}$$

The mechanical dissipation inequality (3.3.6) requires

$$2\nu\mathbf{D}:\mathbf{D}+\lambda(tr\mathbf{D})^2 \geq 0 \tag{3.3.10}$$

since $\mathbf{I}:\mathrm{D} = tr\mathbf{D}$. We can write $\mathbf{D}:\mathbf{D}$ as

$$\mathbf{D}:\mathbf{D} = [\mathbf{D}-\tfrac{1}{3}(tr\mathbf{D})\mathbf{I}]:[\mathbf{D}-\tfrac{1}{3}(tr\mathbf{D})\mathbf{I}]+\tfrac{1}{3}(tr\mathbf{D})^2 \tag{3.3.11}$$

and substituting this expression into (3.3.10) yields

$$2\nu[\mathbf{D}-\tfrac{1}{3}(tr\mathbf{D})\mathbf{I}]:[\mathbf{D}-\tfrac{1}{3}(tr\mathbf{D})\mathbf{I}]+(\lambda+\tfrac{2}{3}\nu)(tr\mathbf{D})^2 \geq 0 \tag{3.3.12}$$

Since (3.3.10) has been reduced to a sum of squares, the necessary and sufficient conditions that will ensure the validity of the inequality are that the coefficient of shear viscosity ν and the bulk or volume viscosity $(\lambda+\frac{2}{3}\nu)$ be positive.

These two examples serve to illustrate an interesting point. We have, in fact, used the Clausius–Duhem inequality as a stability criterion instead of as a restriction on the constitutive relations. The thermodynamic stability conditions are that the thermal conductivity, the bulk and shear viscosity coefficients should all be positive. In other words, we have accepted the validity of the Fourier law (3.3.7) and the Newtonian law (3.3.9) and then used the Clausius–Duhem inequality as a criterion of stability. We see here a blending of the interpretations of the second law that are used in rational and generalised thermodynamics.

The Coleman and Mizel (1964) analysis shows that the free energy determines the entropy but it only determines the elastic part of the stress for a body at rest. Coleman (1964*a*, 1964*b*) considered the question of whether an elastic material can also be dissipative. For a perfectly elastic material we know that equilibrium thermodynamics is applicable, since the free energy depends only on the present state of the system. However, in a more general circumstance, the stress may be influenced by the entire past history of the strain. In this case, will the system be dissipative and if so can the free energy determine the internal dissipation?

In the following, there is a somewhat simplified version of Coleman's analysis (Coleman, 1964*a*) of materials with memory. Moreover, it will be shown that the concept of dissipation is essentially equivalent to that presented above. In order to accomplish this we shall have to rely on Eringen's formulation (Eringen, 1965) of 'smooth' memory as it is applied to rate dependent materials.

In place of the constitutive relation (3.2.30) Coleman, in accordance with the axiom of determinism, assumes that the specific free energy depends not only on the present state of the material but also on its past history.

$$\psi = \hat{\psi}[\theta(t-\tau),\, \mathbf{F}(t-\tau)], \qquad 0 \leq \tau < \infty \tag{3.3.13}$$

where we treat the more general case in which it is necessary to distinguish the functional from its value. We abbreviate the notation by introducing $\mathbf{\Lambda}$ for the ordered pair $(\theta, \mathbf{F})$. $\mathbf{\Lambda}(t-\tau)$ is then decomposed into a part that depends on the present state of the system and a part which depends on its whole past history

$$\mathbf{\Lambda}(t-\tau) = \mathbf{\Lambda}'(t)+\mathbf{\Lambda}_t''(\tau) \tag{3.3.14}$$

The assumption of 'fading memory' is then introduced, which states that the distant history of the system has little influence on its present state. This is made precise by the use of an 'influence' function to define the norm $\|\mathbf{\Lambda}_t''\|$. The influence function is usually taken to be an exponential decreasing function of time. Hence we can write

$$\hat{\psi}[\mathbf{\Lambda}'(t)+\mathbf{\Lambda}_t''(\tau)] = \hat{\psi}[\mathbf{\Lambda}(t)]+\delta\hat{\psi}[\mathbf{\Lambda}'(t)|\mathbf{\Lambda}_t''(\tau)]+0(\|\mathbf{\Lambda}_t''\|), \quad \|\mathbf{\Lambda}_t''\| \to 0 \tag{3.3.15}$$

where $\delta\hat{\psi}[\mathbf{\Lambda}'t)|\mathbf{\Lambda}_t''(\tau)]$ is the Fréchet or 'strong' derivative (Sneddon, 1971). Moreover, it is linear in $\mathbf{\Lambda}_t''(\tau)$. Taking the time derivative (i.e. the 'material' derivative) of (3.3.15) yields

$$\dot{\psi}(t) = \partial_{\mathbf{\Lambda}}\hat{\psi}[\mathbf{\Lambda}(t)]\dot{\mathbf{\Lambda}}(t)+\delta\hat{\psi}[\mathbf{\Lambda}'(t)|\,\dot{\mathbf{\Lambda}}_t''(\tau)] \tag{3.3.16}$$

Then substituting (3.3.16) into the Clausius – Duhem inequality (3.3.3) (with $\mathbf{T}_D = \mathbf{0}$, which obviously excludes the Navier – Stokes fluid) we obtain

$$\gamma - \frac{1}{\rho\theta^2}\boldsymbol{q}\cdot\boldsymbol{g} \geq 0 \tag{3.3.17}$$

where γ is the 'internal' dissipation (Coleman, 1964*a*)

$$\gamma \equiv -\frac{1}{\theta}\delta\hat{\psi}[\mathbf{\Lambda}'(t)|\dot{\mathbf{\Lambda}}_t''(\tau)] \tag{3.3.18}$$

On setting $\boldsymbol{g} = \boldsymbol{0}$ one obtains

$$\gamma \equiv -\frac{1}{\theta}\delta\hat{\psi}[\mathbf{\Lambda}'(t)|\dot{\mathbf{\Lambda}}_t''(\tau)] \geq 0 \tag{3.3.19}$$

which is called the internal dissipation inequality. Observe the striking similarity with the mechanical dissipation inequality (3.3.6). However, in the present case, the claim is that the free energy determines the internal dissipation functional γ.

It is now essential to establish the connection with the thermodynamics of elastic materials with linear viscosity. Coleman (1964*a*) claims that there is no connection. Truesdell (1969) goes further by requiring some 'specialising assumption such as to exclude the Navier – Stokes fluid'. He then proposes the assumption of fading memory. However, Truesdell quickly comes to the conclusion that 'the Navier – Stokes fluid forgets all its past experience almost immediately'. The only way out of this dilemma is to suppose that 'there are many kinds of fading memory'.

We now go on to show that the handling of dissipative effects in the two theories is essentially identical. There is left only the conceptual interpretation of what is meant by dissipation.

Eringen (1965) has also considered constitutive functionals of the form (3.3.13) and has advanced an interpretation of fading memory in terms of 'smooth' memory. For the ordered pair $\mathbf{\Lambda}(t-\tau)$ the assumption of smooth memory allows him to suppose that the function $\mathbf{\Lambda}(t-\tau)$ admits a Taylor series expansion about $\tau = 0$

$$\mathbf{\Lambda}(t-\tau) = \mathbf{\Lambda}(t)-\tau\dot{\mathbf{\Lambda}}(t)+\frac{\tau^2}{2}\ddot{\mathbf{\Lambda}}(t)+\ \ldots \tag{3.3.20}$$

Then the memory of a material at t will be smooth if the constitutive functional $\hat{\mathbf{T}}[\mathbf{\Lambda}(t-\tau)]$ is smooth enough to allow representations of the form

$$\mathbf{T} = \hat{\mathbf{T}}[\mathbf{\Lambda}, \ldots, \mathbf{\Lambda}^{(n)}] \tag{3.3.21}$$

The highest time derivative $\mathbf{A}^{(n)}$ is indicative, in some manner, of the extent of memory. Materials that are described by constitutive functions of the form (3.3.21) are designated as 'materials of the rate type' (Eringen, 1965). Obviously, included in this definition of memory is the Navier – Stokes fluid, albeit it has a short memory, $n = 1$.

Employing constitutive relations of the form (3.3.21), everything follows as before: the rate variables drop out of the constitutive relation for the free energy and the stress splits into elastic and dissipative parts (3.3.2). The mechanical dissipation inequality (3.3.6) is generalised to

$$\gamma \equiv \frac{1}{\theta} tr\{\hat{\mathbf{T}}_D(\mathbf{A}, \ldots, \mathbf{A}^{(n)})\mathbf{D}\} \geq 0 \tag{3.3.22}$$

which provides an explicit form for the internal dissipation (3.3.18).

Since Coleman's procedure, which expresses the history as a first Fréchet derivative (3.3.15), and Eringen's expansion, which expresses it in terms of a Taylor series (3.3.20), both depend upon the assumption that the history has little influence over the present state (one using an influence function and the other considering only small times in the past) and that they are alternative formulations of the same problem. One keeps track of dissipation in terms of an unknown function, the so-called internal dissipation, while the other accounts for dissipation in the dissipative part of the stress. There only remains a conceptual difference as to what is meant by dissipation.

There is still another approach to description of internal dissipation. Coleman and Gurtin (1967) have accounted for dissipative effects in terms of internal state variables or 'hidden' variables. This approach is undoubtedly the closest connection made between rational thermodynamics and the classical formulation of non-equilibrium thermodynamics. In essence, what Coleman and Gurtin have derived is the dissipation function.

The set of internal state variables is designated by the vector $\boldsymbol{a}$

$$\boldsymbol{a} = [\alpha_1, \alpha_2, \ldots, \alpha_N] \tag{3.3.23}$$

These variables are governed by a set of first order kinetic equations

$$\dot{\boldsymbol{a}} = \boldsymbol{f}(\theta, \mathbf{F}, \boldsymbol{g}, \boldsymbol{a}) \tag{3.3.24}$$

Unlike the corresponding set of constitutive relations

$$\psi = \psi(\theta, \mathbf{F}, \boldsymbol{g}, \boldsymbol{a}) \tag{3.3.25}$$

$$\eta = \eta(\theta, \mathbf{F}, \boldsymbol{g}, \boldsymbol{a}) \tag{3.3.26}$$

$$\mathbf{T} = \boldsymbol{T}(\theta, \mathbf{F}, \boldsymbol{g}, \boldsymbol{a}) \tag{3.3.27}$$

$$\boldsymbol{q} = \boldsymbol{q}(\theta, \mathbf{F}, \boldsymbol{g}, \boldsymbol{a}) \tag{3.3.28}$$

the set of kinetic equations (3.3.24) are determined by a specific thermodynamic process and are supposedly known from the start. The total derivative of the specific free energy is

$$\dot{\psi} = \partial_\theta \psi \dot{\theta} + \partial_{\mathbf{F}} \psi : \dot{\mathbf{F}} + \partial_{\boldsymbol{g}} \psi \cdot \dot{\boldsymbol{g}} + \partial_{\boldsymbol{a}} \psi \cdot \dot{\boldsymbol{a}} \tag{3.3.29}$$

where $(\partial_{\boldsymbol{a}}\psi)$ is a vector with N-components

$$\partial_{\boldsymbol{a}}\psi = [\partial_{\alpha_1}\psi, \partial_{\alpha_2}\psi, \ldots, \partial_{\alpha_N}\psi] \tag{3.3.30}$$

By the usual routine procedure, we work our way down to the Clausius–Duhem inequality

$$\partial_{\boldsymbol{a}}\psi\cdot\dot{\boldsymbol{a}}+\frac{1}{\rho\theta}\boldsymbol{q}\cdot\boldsymbol{g}\leq 0 \tag{3.3.31}$$

Observe that the first term in (3.3.31) does not vanish by the usual argument since the internal state variables and their time derivative cannot be assigned arbitrarily due to the connection provided by the kinetic equations (3.3.24).

For an isothermal process $\boldsymbol{g}=\boldsymbol{0}$, we obtain the internal dissipation inequality

$$\partial_{\boldsymbol{a}}\psi\cdot\dot{\boldsymbol{a}}\leq 0 \tag{3.3.32}$$

and the internal dissipation γ is now defined

$$\gamma\equiv-\frac{1}{\theta}\partial_{\boldsymbol{a}}\psi\cdot\dot{\boldsymbol{a}} \tag{3.3.33}$$

Coleman and Gurtin (1967) also observed the identical role of the internal dissipation (3.3.33) and that which has been obtained by the assumption of fading memory (3.3.18). Moreover, since $\boldsymbol{g}$ drops out of the constitutive relation for the free energy (3.3.25), the internal dissipation inequality (3.3.32) and (3.3.29) imply

$$\dot{\psi}\leq 0 \qquad \text{for } \dot{\mathbf{F}}=\mathbf{0} \text{ and } \dot{\theta}=0 \tag{3.3.34}$$

which, expressed in words, says that the specific Helmholtz free energy is always a decreasing function of time for a thermodynamic process with constant temperature and strain. Coleman and Gurtin (1967) then go on to show that equilibrium is characterised by that state of lowest free energy for a thermodynamic process with constant strain and temperature. This is to be considered as a fundamental criterion of stability.

However, a little consideration of classical non-equilibrium thermodynamics will quickly reveal that what these authors have derived is none other than the dissipation function ϕ (Landau and Lifshitz, 1959)

$$2\phi\equiv-\frac{1}{\theta}\dot{\psi} \tag{3.3.35}$$

when $\dot{\mathbf{F}}=\mathbf{0}$ and $\dot{\theta}=0$. In this sense, the free energy determines the dissipation function. In passing, it is interesting to refer to a comment made by Truesdell (1969) in which he concludes that, on account of the apparent difficulty with the Onsager symmetry relations of near equilibrium thermodynamics, 'little is said about dissipation functions in more recent Onsagerist literature'. Yet, when we consider rational thermodynamics, we find that this is their major accomplishment: they have obtained alternative forms for the dissipation function, cf. (3.3.18), (3.3.22) and (3.3.33).

From the foregoing discussion it would appear that $\phi\geqq 0$ has the same information content as the second law when $\boldsymbol{g}=\boldsymbol{0}$. A little consideration shows this not to be the case. We generalise the energy balance equation (3.1.3) by the addition of an external power supply

$$\dot{\varepsilon}+\frac{1}{\rho}\nabla\cdot\boldsymbol{q}=\frac{1}{\rho}tr\{\mathbf{T}\,\mathbf{L}\}+\theta\pi \tag{3.3.36}$$

where π is the power per unit mass and temperature. We dropped the mysterious radiation energy term, since we do not need it for self-consistency reasons. Eliminating the divergence of the heat flux vector between the entropy and internal energy balance equations, we obtain a power balance relation in place of the expected Clausius – Duhem inequality

$$\theta(\pi-\sigma_s)=\dot{\psi}+\eta\dot{\theta}+\frac{1}{\rho\theta}\boldsymbol{q}\cdot\boldsymbol{g}-\frac{1}{\rho}tr\{\mathbf{T}\,\mathbf{L}\} \tag{3.3.37}$$

The free energy certainly does not determine the power and there is no justification for eliminating it from equation (3.3.37). Moreover, we have no means by which *a priori* bounds can be imposed on the magnitude of the power although we know it to have a negative lower bound, since the body is finite. Consequently, we cannot interpret (3.3.37) as an inequality for an arbitrary thermodynamic process, since there is a factor that can be negative as well as positive, that is the rate at which work is being done by or on the system. We are therefore constrained to view (3.3.37) as a power balance principle rather than an inequality. A great deal more is said about the power balance principle in chapter 5.

3.4 Limitations of rational thermodynamics

The basic idea underlying rational thermodynamics is the employment of the second law as a restriction on the form of the constitutive equations. No one ever talks about violations of the Clausius – Duhem inequality and their relation to non-equilibrium instabilities. All stability criteria are formulated in terms of the extrema of thermodynamic potentials, using the Clausius – Duhem inequality (Erickson, 1966). No conditions for instability have ever been formulated in rational thermodynamics. Moreover, we know that the stability criteria, expressed in terms of the extrema of thermodynamic potentials, is only valid in the immediate neighbourhood of equilibrium. This provides a clue in the determination of the limits of rational thermodynamics. Nevertheless, rational thermodynamics has been used in the analysis of the onset and propagation of thermal waves in stationary rigid conductors (Bogy and Naghdi, 1970). We shall now show that the condition for the onset of thermal wave propagation arises as the result of instability in which the Clausius – Duhem inequality is violated.

Since the material is a rigid heat conductor, we start with the constitutive relations

$$\psi=\psi(\theta,\ \boldsymbol{g}) \tag{3.4.1}$$

$$\eta=\eta(\theta,\ \boldsymbol{g}) \tag{3.4.2}$$

$$\boldsymbol{q}=\boldsymbol{q}(\theta,\ \boldsymbol{g}) \tag{3.4.3}$$

Since there is no stress, the Clausius – Duhem inequality reduces to the heat conduction inequality (3.2.34). The energy balance equation is

$$\rho\dot{\varepsilon}+\nabla\cdot\boldsymbol{q}=\rho r \tag{3.4.4}$$

Moreover, since the constitutive relations are given in the form of (3.4.1–3) it proves convenient to replace the specific internal energy in (3.4.4) by the specific Helmholtz free energy, cf. (3.2.12),

$$\rho(\dot{\psi}+\eta\dot{\theta}+\dot{\eta}\theta) = -\nabla\cdot\boldsymbol{q}+\rho r \tag{3.4.5}$$

which is to be treated as the equation for heat conduction. Since ψ is independent of $\boldsymbol{g}$ and

$$\frac{\partial}{\partial\theta}\psi(\theta) = -\eta(\theta) \tag{3.4.6}$$

the heat conduction equation can be written in the form

$$\rho\theta\left(\frac{\partial^2\psi}{\partial\theta^2}\right)\dot{\theta} = +\rho r-\left(\frac{\partial q}{\partial\theta}\right)\theta_{,x} -\left(\frac{\partial q}{\partial g}\right)\theta_{,xx} \tag{3.4.7}$$

where, for simplicity, we have considered the one-dimensional case: q and g are the x components of $\boldsymbol{q}$ and $\boldsymbol{g}$, respectively, and the comma denotes partial differentiation with respect to the spatial coordinate. Equation (3.4.7) is of the parabolic type with $(\partial^2\psi/\partial\theta^2) > 0$ and $(\partial q/\partial g) < 0$. No propagation of thermal waves can occur. This is also evident from the Clausius–Duhem inequality; the entropy produced is due solely to heat conduction. In order to obtain propagating thermal waves, it is necessary to generalise the constitutive relations (3.4.1–3) to include the temperature rate.

Bogy and Naghdi now assume that the constitutive relations are of the form

$$\psi = \psi(\theta, g, \dot{\theta}) \tag{3.4.8}$$

$$\eta = \eta(\theta, g, \dot{\theta}) \tag{3.4.9}$$

$$q = q(\theta, g, \dot{\theta}) \tag{3.4.10}$$

By the same line of reasoning used in section 3.2, it is found that $\dot{\theta}$ and g drop out of the free energy constitutive relation and the entropy production (i.e. the Clausius–Duhem inequality) is

$$\theta\sigma_S = -\left(\frac{\partial\psi}{\partial\theta}+\eta\right)\dot{\theta}-\frac{1}{\rho\theta}q\cdot g \geq 0 \tag{3.4.11}$$

Nothing has been gained with respect to the previous case if the free energy determines the entropy completely, cf. (3.4.6). However, no restrictions have been imposed on the constitutive relations (3.4.9, 10) so that one could suppose that the entropy is made up of two parts in a similar way that the stress (3.3.2) is comprised of elastic and dissipative parts. Thus we write

$$\eta(\theta, g, \dot{\theta}) = \eta_0(\theta)+\eta_I(\theta, g, \dot{\theta}) \tag{3.4.12}$$

Eringen (1975) has called $\eta_I(\theta, g, \dot{\theta})$ the 'dissipative' entropy, in analogy with the dissipative part of the stress $\mathbf{T}_D$. However, this analogy is misleading since, as we shall shortly see, cf. (3.4.14), $\eta_I(\theta, g, \dot{\theta})$ is by no means dissipative even though it is a function of the temperature rate. The free energy still determines the 'equilibrium' part of the entropy

$$\frac{\partial}{\partial\theta}\psi(\theta) = -\eta_0(\theta) \tag{3.4.13}$$

and, consequently, the Clausius–Duhem inequality (3.4.11) reduces to

$$\theta\sigma_S = -\eta_I(\theta, g, \dot{\theta})\dot{\theta} - \frac{1}{\rho\theta}q(\theta, g, \dot{\theta})\cdot g \geq 0 \tag{3.4.14}$$

Now, not only does heat conduction contribute to the production of entropy but there is also a source term. It is this source term that is to be implicated in the cause of instability and the onset of thermal wave propagation, as the following analysis will clearly show.

Suppose that there exists a time independent stationary state which is designated by θ_0 and g_0. Considering small perturbations about this state, we write

$$\theta(t) = \theta_0 + \xi_1(t) \tag{3.4.15}$$

$$g(t) = g_0 + \xi_2(t) \tag{3.4.16}$$

such that the perturbations $\xi_1(t)$ and $\xi_2(t)$ satisfy $|\xi_1(t)/\theta(t)|, |\xi_2(t)/g(t)| \ll 1$. Expanding the functions $\eta_I(\theta, g, \dot{\theta})$ and $q(\theta, g, \dot{\theta})$ to first order in the perturbation variables, we obtain

$$\eta_I(\theta, g, \dot{\theta}) = \eta_I^0(\theta_0, g_0) + \partial_\theta \eta_I \xi_1 + \partial_g \eta_I \xi_2 + \partial_{\dot{\theta}} \eta_I \dot{\xi}_1 \tag{3.4.17}$$

$$q(\theta, g, \dot{\theta}) = q^0(\theta_0, g_0) + \partial_\theta q \xi_1 + \partial_g q \xi_2 + \partial_{\dot{\theta}} q \dot{\xi}_1 \tag{3.4.18}$$

where all partial derivatives are to be evaluated at the stationary state θ_0, g_0. Substituting (3.4.17) and (3.4.18) into the Clausius–Duhem inequality (3.4.14) leads to

$$\begin{aligned}
&\frac{1}{\rho\theta_0}q^0 g_0 + \frac{g_0}{\rho\theta_0}\partial_\theta q \xi_1 + \frac{1}{\rho\theta_0}(q^0 + g_0 \partial_g q)\xi_2 + \left[\frac{1}{\rho\theta_0}(g_0 \partial_{\dot{\theta}} q) + \eta_I^0\right]\dot{\xi}_1 \\
&+ \frac{1}{\rho\theta_0}\partial_\theta q \xi_1 \xi_2 + \left[\frac{1}{\rho\theta_0}\partial_{\dot{\theta}} q + \partial_g \eta_I\right]\dot{\xi}_1 \xi_2 + \partial_\theta \eta_I \dot{\xi}_1 \xi_2 + \frac{1}{\rho\theta_0}\partial_g q \xi_2^2 \\
&+ \partial_{\dot{\theta}} \eta_I \dot{\xi}_1^2 \leq 0
\end{aligned} \tag{3.4.19}$$

Stationary state stability requires ($\xi_1 = \xi_2 = 0$)

$$q^0(\theta_0, g_0)g_0 \leq 0 \tag{3.4.20}$$

which is just the heat conduction inequality (3.2.34), evaluated at the stationary state.

As a simplified situation, let us suppose that the stationary value of the temperature gradient vanishes, $g_0 = 0$ and that the θ dependencies of q and η_I are negligibly small. Then the necessary and sufficient conditions that the entropy production be positive definite are

$$\left(\frac{1}{\rho\theta_0}\partial_g q + \partial_{\dot{\theta}} \eta_I\right) < 0 \tag{3.4.21}$$

and

$$(\partial_{\dot{\theta}} q + \rho\theta_0 \partial_g \eta_I)^2 - 4\rho\theta_0(\partial_g q)(\partial_{\dot{\theta}} \eta_I) < 0 \tag{3.4.22}$$

We now compare the criteria of stability with the stability condition obtained from the heat conduction equation.

Under the same conditions by which we derived the stability conditions (3.4.21, 22), the linearised heat conduction equation is

$$\rho\theta_0(\partial_{\dot\theta}\eta_I)\ddot\theta + [\partial_{\dot\theta} q + \rho\theta_0(\partial_g\eta_I)]\theta_{,x} + (\partial_g q)\theta_{,xx} = 0 \tag{3.4.23}$$

We assume that equation (3.4.23) admits a stationary propagating wave solution with an assumed phase speed U. If such a solution does, in fact, exist then the expression for U must be real:

$$U = \frac{1}{2\rho\theta_0(\partial_{\dot\theta}\eta_I)}\{\partial_{\dot\theta} q + \rho\theta_0\partial_g\eta_I \pm [(\partial_{\dot\theta} q + \rho\theta_0\partial_g\eta_I)^2 - 4\rho\theta_0(\partial_g q)(\partial_{\dot\theta}\eta_I)]^{1/2}\} \tag{3.4.24}$$

A comparison of condition (3.4.22) and the phase speed expression (3.4.24) quickly reveals that the onset of thermal wave propagation coincides with the instability of the stationary state. This is indeed a logical result, since the transition from the stationary state to a state of stationary propagating waves must be caused by some internal mechanism. The cause of the transition is the instability of the stationary state, as we have deduced from the criterion of the second law. But notice that we have inverted the rational thermodynamic interpretation of the second law. Instead of using the criterion of the second law as a restriction on the form of the constitutive equations, we have asked for the conditions under which it may no longer be valid. This is to say, we have employed the second law as a criterion of stability rather than a restriction on the form of the constitutive relations. This illustration reveals the basic distinction between the rational thermodynamic and generalised thermodynamic interpretation of the second law.

References

Bogy, D. B. and Naghdi, P. M. (1970). On heat conduction and wave propagation in rigid solids. *J. math. Phys.*, **11**, 917–923.

Coleman, B. D. (1964*a*). Thermodynamics of materials with memory. *Archs rational Mech. Analysis*, **17**, 1–46.

Coleman, B. D. (1964*b*). On thermodynamics, strain impulses, and visco-elasticity. *Archs rational Mech. Analysis*, **17**, 230–254.

Coleman, B. D. and Gurtin, M. E. (1967). Thermodynamics with internal state variables. *J. chem. Phys.*, **47**, 597–613.

Coleman, B. D. and Mizel, V. J. (1964). Existence of caloric equations of state in thermodynamics. *J. chem. Phys.*, **40**, 1116–1125.

Coleman, B. D. and Noll, W. (1962). Simple fluids with fading memory. Proc. Internat. Symp. on. Second-Order Effects, Haifa, pp. 530–552. Macmillan, New York.

Eckart, C. (1940). Thermodynamics of irreversible processes I. The simple fluid. *Phys. Rev.*, **58**, 267–269; II. Fluid mixtures. *Phys. Rev.*, **58**, 269–275.

Edelen, D. G. B. and McLennan, J. A. (1973). Material indifference: a principle or a convenience. *Int. J. engng Sci.*, **11**, 917–923.

Erickson, J. L. (1966). Thermoelastic stability. *Proc. 5th US natn. Congr. appl. Mech.*, p. 187–193.

Eringen, A. C. (1965). A unified theory of thermomechanical materials. *Int. J. engng Sci.*, **4,** 179–202.

Eringen, A. C. (1975). *Continuum Physics*, p. 89 (Ed. Eringen, A. C.). Academic Press, New York.

Gurtin, M. E. (1968). On the thermodynamics of materials with memory. *Archs rational Mech. Analysis*, **28,** 40–50.

Landau, L. D. and Lifshitz, E. M. (1959). *Statistical Physics*, p. 378. Pergamon Press, Oxford.

Müller, I. (1967). On the entropy inequality. *Archs rational Mech. Analysis*, **26,** 118–141.

Prigogine, I. (1949). Le domaine de validité de la thermodynamique des phénomènes irréversibles. *Physica*, **15,** 272–284.

Sneddon, I. N. (1971). *Continuum Physics*, p. 447 (Ed. Eringen, A. C.). Academic Press, New York.

Truesdell, C. (1966). *Six Lectures on Modern Natural Philosophy*. Springer-Verlag, Berlin and Heidelberg.

Truesdell, C. (1969). *Rational Thermodynamics*. McGraw-Hill, New York.

Truesdell, C. and Noll, W. (1965). The classical field theories, in *Handbuch der Physik*, vol. III/3, p. 294 (Ed. Flügge, S.), Springer-Verlag, Berlin.

Truesdell, C. and Toupin, R. A. (1960). The classical field theories, in *Encyclopedia of Physics*, vol. III/1, pp. 703–704 (Ed. Flügge, S.), Springer-Verlag, Berlin.

4 'Generalised' Thermodynamics

Another school of non-equilibrium thermodynamics has been based on an approach which extends equilibrium concepts to non-equilibrium thermodynamics. We call this approach 'generalised' thermodynamics, although it should not be confused with Tisza's use (Tisza, 1966) of the term in an equilibrium context. The central dogma of the thermodynamic theory is to be found in the work of Glansdorff and Prigogine (1971). A rather succinct presentation of the theory, together with interesting, although thermodynamically unrelated, examples of non-equilibrium processes can be found in the review of Nicolis (1971). More recently the emphasis has been shifted from a completely thermodynamic to a kinetic justification of the Glansdorff–Prigogine thermodynamic stability criteria (Glansdorff, Nicolis and Prigogine, 1974).

Unlike rational thermodynamics, generalised thermodynamics has sought justification for its approach using microscopic or kinetic theory arguments. Prigogine (1949) has shown that the classic Gibbs equation is valid to first order in the perturbed distribution function that was calculated from the Boltzmann equation using the Chapman–Enskog procedure. Therefore, for small deviations in the distribution function from its equilibrium value, we may still expect equilibrium thermodynamics to apply. This finding has been formalised into the fundamental postulate of generalised thermodynamics or what Landsberg (1972) refers to as the 'basic trick'. It is more often referred to as the 'local equilibrium' assumption.

The system is broken up into a number of small elements which are large enough so that it makes sense to define thermodynamic potentials, yet small enough so that in each element of volume the gradients can be considered negligible. The composite system may not be in equilibrium, since the chemical potentials, temperatures, etc., may vary from element to element. Local equilibrium implies that the specific entropy η is the same function of extensive variables per unit mass as it is at equilibrium.

Moreover, the assumption of a 'stable' local equilibrium, according to Glansdorff and Prigogine (1971), implies that the entropy surface is convex

$$\delta^2\eta = \partial_{\alpha_i}\partial_{\alpha_j}\eta\xi_i\xi_j \leq 0 \tag{4.1}$$

However, no condition is imposed on the first variation of the entropy, $\delta\eta$, so that we do not know if each element is in a true state of equilibrium with maximum entropy or not. When condition (4.1) is coupled with

$$\tfrac{1}{2}\partial_t\delta^2\eta \geq 0 \tag{4.2}$$

we have the stability criteria put forward by the school of generalised thermodynamics. Furthermore, for small deviations from equilibrium, (4.2) will coincide with the specific entropy production whose positive semidefinite form is guaranteed by the second law. On the contrary, for small deviations from a non-equilibrium stationary state, neither $\delta\eta$ nor $\partial_t\delta\eta$ vanishes and inequality (4.2) has no connection with the second law. It is then to be inferred that (4.2) is a separate 'thermodynamic principle' (Landsberg, 1972), with an information content all of its own. The application of the thermodynamic criterion (4.2) has been criticised by Keizer and Fox (1974).

Prior to a discussion of the validity of the thermodynamic criteria (4.1) and (4.2), which is undertaken in section 4.2, it is necessary to develop the basic ideas leading to the thermodynamic criteria. Here, we do not have a set of well-defined axioms as we have in rational thermodynamics. Criteria (4.1) and (4.2) have to be understood in a historical context, since they are the culmination of over three decades of work in generalised thermodynamics. Although we shall encounter fundamental and inherent difficulties in the thermodynamic theory of stability, it is only fitting to accredit the school of generalised thermodynamics with (1) the use of the second law as a criterion of stability, and (2) the clear distinction between near-equilibrium behaviour as opposed to the dynamic behaviour of non-equilibrium processes far from equilibrium. In particular, interest has been focused on non-equilibrium instabilities and the phenomena accompanying transitions in the state of the system.

A great deal of care must be exercised in distinguishing between linear and nonlinear thermodynamics and linear and nonlinear processes. Generalised thermodynamics is a *linear* theory; it treats small deviations from equilibrium or non-equilibrium stationary states. Linear thermodynamics is applicable to small deviations from equilibrium, whereas nonlinear thermodynamics may (or may not) be applicable to the analysis of small motions about non-equilibrium stationary states. It is wrong and misleading to conclude that the stability criteria of linear thermodynamics are applicable to 'steady states that belong to the linear range' (Prigogine and Lefever, 1975). There is only one steady state in the linear region and that is the state of equilibrium. On the other hand, linear and nonlinear processes refer to the linearity or nonlinearity of the phenomenological relations that connect the velocities to their 'causes'—the thermodynamic forces. Generalised thermodynamics does not treat nonlinear processes.

Care must also be exercised in the distinction between thermodynamic and kinetic criteria of stability. To justify a thermodynamic criterion on kinetic criteria of stability does not make the criterion thermodynamic. In the more recent literature on generalised thermodynamics (Glansdorff, Nicolis and Prigogine, 1974), this distinction becomes more and more nebulous, since it has been found that non-equilibrium thermodynamics has proved incapable of establishing the criterion (4.2) as a general thermodynamic principle.

The method of generalised thermodynamics is to assume the validity of the

Gibbs equation and to ask for the thermodynamic conditions for stability or instability. In addition to the criteria (4.1) and (4.2) there exist a whole host of other 'thermodynamic' criteria for which we are unable to provide a logical structure of generalised thermodynamics, as we have done for rational thermodynamics. There is no other alternative but to give a historical development of generalised thermodynamics.

4.1 The development of generalised thermodynamics

Generalised thermodynamics recognises the fact that thermodynamic potentials *only* possess free extrema at equilibrium. This is due to the fact that the first variations of the thermodynamic potentials do not vanish for non-equilibrium stationary states. It is necessary that there is a finite value of the thermodynamic force at the stationary state which prevents the system from relaxing back to equilibrium. Consequently, one of the major efforts of generalised thermodynamics was to search for non-equilibrium variational principles that would characterise the stability of non-equilibrium stationary states in an analogous way such that the extrema of equilibrium thermodynamic potentials determine the stability of the equilibrium state.

In an apparent analogy with Onsager's variational principle of isolated, non-equilibrium processes (cf. section 6.2), Prigogine (1947, 1954) raised the question as to whether the entropy production could serve as a non-equilibrium thermodynamic potential for non-equilibrium stationary states. His result has been formulated as the 'theorem of minimum production of entropy'. However, as we shall see, the entropy production does not satisify a variational principle, or even a minimum principle, and actually what Prigogine derived was the principle of least dissipation of energy applied to small deviations from equilibrium. However, in section 6.2, it is shown that the principle of minimum entropy production is a by-product of Onsager's variational principle. Prior to a discussion of Prigogine's demonstration, we have to discuss the formulation of generalised thermodynamics.

In generalised thermodynamics it is customary to decompose the entropy into two parts

$$\mathrm{d}S = \mathrm{d}_e S + \mathrm{d}_i S \tag{4.1.1}$$

where $\mathrm{d}_e S$ is the entropy exchanged between system and surroundings and $\mathrm{d}_i S$ is the entropy produced in the system. Equation (4.1.1) is just another way of writing the entropy balance relation. In order to identify these two terms, we employ the Gibbs equation

$$\mathrm{d}S = \frac{1}{\theta}\mathrm{d}E + \frac{1}{\theta}p\mathrm{d}V - \boldsymbol{X}\cdot\mathrm{d}\boldsymbol{a} \tag{4.1.2}$$

where the thermodynamic force vector is defined

$$\partial_{\boldsymbol{a}} S = -\boldsymbol{X}. \tag{4.1.3}$$

The use of vector notation simplifies the use of having to write sums of terms. The Gibbs equation (4.1.2) is assumed to be valid on the basis of the local equilibrium postulate. Moreover, the postulate of local equilibrium prohibits the treatment of the class of phenomena encountered in rational thermodynamics. The Gibbs equation (4.1.2) is identical to the equilibrium expression and we cannot add additional generalised work terms, since each element of the system is assumed to be in a state of local equilibrium.

Inserting the first law,

$$dE = dQ - p\,dV \tag{4.1.4}$$

into the Gibbs equation (4.1.2) yields

$$dS = \frac{1}{\theta}dQ - \boldsymbol{X}\cdot d\boldsymbol{\alpha} \tag{4.1.5}$$

which is now compared with (4.1.1). We find

$$d_eS = \frac{1}{\theta}dQ \tag{4.1.6}$$

and

$$d_iS = -\boldsymbol{X}\cdot d\boldsymbol{\alpha} \tag{4.1.7}$$

Dividing (4.1.6) formally by dt and then by the constant total mass, we obtain the expression for the specific entropy production for an isothermal process

$$\sigma_S = -\boldsymbol{X}\cdot\dot{\boldsymbol{\alpha}} \equiv \boldsymbol{X}\cdot\boldsymbol{\iota} \tag{4.1.8}$$

We now make use of the fact that the velocity vector $\boldsymbol{\iota}$ is related to the internal state variable vector $\boldsymbol{\alpha}$ by a set of kinetic equations

$$\boldsymbol{\iota} = \boldsymbol{f}(\boldsymbol{\alpha}) \tag{4.1.9}$$

When these equations are linearised about a time independent state and the thermodynamic force variation χ is introduced through an equation of state, cf. equations (4.2.11) and (4.2.13), a set of linear phenomenological relations is obtained

$$\boldsymbol{\chi} = -\boldsymbol{\Lambda}\cdot\boldsymbol{\iota} \tag{4.1.10}$$

where we have chosen the velocities such that their stationary values ι_j^0 vanish. Moreover, if the linearisation has been performed about the state of equilibrium, the Onsager symmetry relations guarantee that the matrix of differential coefficients, that has been evaluated at equilibrium, is symmetric:

$$\boldsymbol{\Lambda} = \boldsymbol{\Lambda}^T \tag{4.1.11}$$

It cannot be overemphasised that there is nothing whatever that will guarantee the symmetry of the phenomenological matrix for any state other than equilibrium. The symmetry is a manifestation of detailed balancing at equilibrium. Other states may have the symmetry indicated in (4.1.11), and when they do we have scalar thermodynamic potentials. This important fact has been played down in generalised thermodynamics and accounts for many of the inherent shortcomings of the theory.

It is a common practice in generalised thermodynamics to introduce the phenomenological relations (4.1.10) into the definition of the entropy production (4.1.8) and write

$$\sigma_S(\iota, \iota) = \boldsymbol{\iota}^T \cdot \boldsymbol{\Lambda} \cdot \boldsymbol{\iota} \tag{4.1.12}$$

or

$$\sigma_S(\chi, \chi) = \boldsymbol{\chi}^T \cdot \boldsymbol{\Lambda}^{-1} \cdot \boldsymbol{\chi} \tag{4.1.13}$$

This leaves open the possibility of expressing the specific entropy production as a quadratic function of the velocities or the forces. Although it is indeed true that (4.1.12) and (4.1.13) are numerically equivalent for isolated, isothermal non-equilibrium processes, they *cannot* be taken as definitions of functional dependencies! The functional dependence of the entropy production is given by (4.1.8): it is a *linear* function of the velocities. When the linear transformation (4.1.10) is introduced into (4.1.8), the correct form of (4.1.13) becomes

$$\sigma'_S(\chi) = \boldsymbol{\chi}^T \cdot \boldsymbol{\Lambda}^{-1} \cdot \boldsymbol{\chi} = \sigma_S(\iota) \tag{4.1.14}$$

Actually, (4.1.12) and (4.1.13) are definitions of the dissipation function and generating function, respectively (cf. sections 5.2 and 5.3). This explains how Prigogine created a minimum principle for the entropy production. Rather, it should have been entitled the free minimum principle of least dissipation of energy, which had already been derived by Onsager, 15 years earlier and in a more general form.

Prigogine (1954) formulated his principle of minimum entropy production in analogy with Gauss's principle of least constraint

$$Z = (\boldsymbol{F} - m\boldsymbol{A})^2 = \min \tag{4.1.15}$$

where $\boldsymbol{F}$ is the force, m is the mass and $\boldsymbol{A}$ is the acceleration. In classical mechanics, (4.1.15) acts as a constraint on the motion. For the actual motion, the constraint becomes as small as possible. Gauss's principle is not a true variational principle, in that it involves the minimisation of a certain function and does not consider the motion as a whole (cf. section 6.3). According to Prigogine we are free to choose either (4.1.12) or (4.1.13). Choosing (4.1.12) we attempt to cast it in the form of Gauss's principle (4.1.15). But we have to assume, as did Prigogine (1954), that the generalised resistance coefficients Λ_{ii} are all equal. Then

$$\sigma_S \sim \iota^2 = \min \tag{4.1.16}$$

acts as a constraint on the evolution of non-equilibrium processes in the same way that Gauss's principle (4.1.15) acts as a constraint on the motion in classical mechanics. Without disputing the truth content of (4.1.16), let us see where it would be applicable.

Nicolis (1971) has argued that the principle of minimum entropy production, as formulated by Prigogine, can be used to characterise non-equilibrium stationary states for which the velocities can be chosen such that their stationary values vanish. The requirements for (4.1.12) to be a minimum are

$$\partial_{\iota_i} \sigma_S = 2\Lambda_{ij}\iota_j = 0 \tag{4.1.17}$$

and

$$\partial^2_{\iota_i} \sigma_S = 2\Lambda_{ii} \geq 0 \tag{4.1.18}$$

Condition (4.1.17) follows from the fact that we have required the velocities to vanish in the stationary state.

However, a little consideration shows this argument to be completely fallacious. The mistake was to introduce the linearised phenomenological relations (4.1.10) into the entropy production (4.1.8). It necessarily supposed that the stationary values of the thermodynamic forces χ_i^0 vanished — which is only true if the stationary state is equilibrium. Under the condition that the stationary value of the velocities vanish, the correct form of the entropy production for states other than equilibrium is

$$\sigma_S = -\boldsymbol{\chi}^0 \cdot \boldsymbol{\imath} - \boldsymbol{\chi} \cdot \boldsymbol{\imath} \tag{4.1.19}$$

where if we want to incorrectly introduce the phenomenological relations (4.1.10) we obtain

$$\sigma_S = -\boldsymbol{\chi}^0 \cdot \boldsymbol{\imath} + \boldsymbol{\imath}^T \boldsymbol{\Lambda} \boldsymbol{\imath}$$

This expression clearly shows that the first variation of the entropy production does *not* vanish for non-equilibrium stationary states. Consequently, Prigogine's formulation of the principle of minimum entropy production is valid only for small displacements from equilibrium. Moreover, there is no guarantee that conditions (4.1.18) would be fulfilled for non-equilibrium stationary states. The Onsager symmetry may no longer be valid and the phenomenological coefficients Λ_{ii} may even be complex!

This has served to illustrate a very important fact of non-equilibrium thermodynamics. Thermodynamic stability criteria cannot be derived from any single non-equilibrium thermodynamic potential, since the first variation of the potential does not vanish in non-equilibrium stationary states. For this reason, the thermodynamic stability criterion (4.2) constitutes a separate thermodynamic principle whose only means for justification must ultimately come from kinetic stability analysis. We shall discuss this point in much greater detail in section 4.2.

A subsequent development in generalised thermodynamics was the Glansdorff–Prigogine (1954, 1964) derivation of their 'universal criterion of evolution'. The motivation for the derivation of the criterion is very simple: they wanted to be able to characterise the evolution of non-equilibrium processes in an analogous way that, for example, the free energy determines the evolution to equilibrium. For isothermal and isochoric processes, the Helmholtz free energy is a decreasing function of time; equilibrium is characterised as that state with the least free energy. Arguing that the entropy production plays an analogous role in non-equilibrium thermodynamics, Glansdorff and Prigogine (1954) set out to determine the time derivative of the entropy production, $\dot{\sigma}_S$.

The differential of the entropy production was split into two parts, one due to a change in the velocity and the other due to a change in the force

$$\mathrm{d}\sigma_S = \mathrm{d}_\imath \sigma_S + \mathrm{d}_X \sigma_S \tag{4.1.20}$$

where

$$\mathrm{d}_\imath \sigma_S = -\boldsymbol{X} \cdot \mathrm{d}\boldsymbol{\imath} \tag{4.1.21}$$

$$\mathrm{d}_X \sigma_S = -\boldsymbol{\imath} \cdot \mathrm{d}\boldsymbol{X} \tag{4.1.22}$$

The first case to be analysed was the linear range where the Onsager symmetry is guaranteed to hold. Changes in the velocity and the force are not independent but rather related by the phenomenological relations (4.1.10). Here, we need not be concerned with the stationary value of the force, since it vanishes at equilibrium; hence $\boldsymbol{X} = \boldsymbol{\chi}$. Introducing the phenomenological relations (4.1.10) into (4.1.21) and (4.1.22) results in

$$\tfrac{1}{2}\mathrm{d}\sigma_S = \Lambda_{ij}\iota_i \mathrm{d}\iota_j = \mathrm{d}_{\iota}\sigma_S = \mathrm{d}_{\chi}\sigma_S \leq 0 \tag{4.1.23}$$

from which it follows that

$$\dot{\sigma}_S \leq 0 \tag{4.1.24}$$

Inequality (4.1.24), expressed in words, states that the entropy production is a decreasing function of time for all stable processes. Equilibrium is characterised by the least production of entropy where it vanishes on account of the fact that it is a positive semidefinite bilinear expression.

The second case that Prigogine and Glansdorff analysed was the nonlinear range where there is no guarantee that the Onsager symmetry holds. For this reason (and not because of an increase in generality) reference is no longer made to the phenomenological equations. The expression for the rate of change of the entropy production is

$$\dot{\sigma}_S = -(\boldsymbol{X}\cdot\dot{\boldsymbol{\iota}} + \boldsymbol{\iota}\cdot\dot{\boldsymbol{X}}) \tag{4.1.25}$$

In spite of all the attempts made (Mel, 1954), no definite sign could be attached to the ${}_{\iota}\dot{\sigma}_S$ term so that Glansdorff and Prigogine (1954, 1964) were content to call

$${}_{\chi}\dot{\sigma}_S = -\boldsymbol{\iota}\cdot\boldsymbol{X} \leq 0 \tag{4.1.26}$$

a 'universal criterion of evolution'. The demonstration of inequality (4.1.26) involved the use of an equation of state which reduced (4.1.26) to the so-called equilibrium criterion of 'stability with respect to diffusion' (de Groot and Mazur, 1962; Prigogine, 1967). We shall derive the inequality under somewhat more general conditions that employ only the convex property of the entropy surface in Gibbs space.

The thermodynamic force is defined by (4.1.3) which in terms of specific quantities becomes

$$\partial_{\boldsymbol{\alpha}}\eta(\varepsilon, \alpha, \rho) = -\boldsymbol{X}(\varepsilon, \alpha, \rho) \tag{4.1.27}$$

Taking the time derivative of (4.1.27) and interchanging the order of the differential operators, we obtain

$$\partial_{\boldsymbol{\alpha}}\dot{\eta}(\varepsilon, \alpha, \rho) = -\dot{\boldsymbol{X}}(\varepsilon, \alpha, \rho) \tag{4.1.28}$$

Since each element of the system is in a state of local equilibrium, that is an isolated system, $\varepsilon = \text{const}$ and $\rho = \text{const}$, so that the total derivative of the specific entropy is $\partial_{\boldsymbol{\alpha}}\eta\cdot\boldsymbol{\iota}$ which, when substituted into (4.1.28), gives

$$\partial_{\boldsymbol{\alpha}}[\partial_{\boldsymbol{\alpha}}\eta\cdot\boldsymbol{\iota}] = -\dot{\boldsymbol{X}} \tag{4.1.29}$$

Provided there is no connection between the velocities and the internal state variables, $\boldsymbol{\iota} \neq f(\boldsymbol{\alpha})$, relation (4.1.29) becomes

$$\partial_{\boldsymbol{\alpha}}^2\eta\boldsymbol{\iota} = -\dot{\boldsymbol{X}} \tag{4.1.30}$$

Then the introduction of (4.1.30) into (4.1.26) leads to the desired result

$$_X\dot{\sigma}_S = \partial_\alpha^2 \eta \iota^2 \leq 0 \tag{4.1.31}$$

which follows from the local equilibrium stability criterion $\partial_\alpha^2 \eta \leq 0$, cf. criterion (4.1). The reason for calling (4.1.31) a universal criterion is that no reference is made to the deviations in the velocities from their stationary state values. The comparison is often drawn between the facts that $d_i S$ and $d_X \sigma_S$ are both non-total differentials (Glansdorff and Prigogine, 1954, 1964). Although this is intended to accentuate the importance of (4.1.31) in view of the second law, the decomposition of the entropy (4.1.1) has no mathematical significance.

The significance of the universal criterion and the manner in which it is derived deserves some comment. It is often claimed (de Groot and Mazur, 1962) that (4.1.31) is much more general than the linear range result (4.1.23), since it makes no reference to the phenomenological relations. If indeed (4.1.31) is a more general result then it also must be true that

$$_X\dot{\sigma}_S = [\Lambda^{-1}]_{ij} \chi_i \dot{\chi}_j \leq 0 \tag{4.1.32}$$

in the less general case when the phenomenological relations (4.1.10) are used. For non-equilibrium stationary states, we recall that the Onsager symmetry is no longer guaranteed, so it is *a priori* impossible to establish a given sign for (4.1.32). Moreover, there is absolutely no connection between the *static* equilibrium criterion $\partial_\alpha^2 \eta \leqq 0$ and the *kinetic* criterion that all the eigenvalues of matrix of differential coefficients, obtained by linearising the kinetic equations (4.1.9) about a particular stationary state, must all have negative real parts.

Consider now the expression for the total differential of the specific entropy production (4.1.20). For $N+1$ components we can write it in the form

$$d\sigma_S = -\chi_j d\iota_j - \iota_j d\chi_j \tag{4.1.33}$$

where the sum extends over all the $N+1$ components. We ask for the conditions that will determine whether or not $d\sigma_S$ is a complete or exact differential (Lavenda, 1972). From (4.1.33) we find that there are three sets of integrability conditions:

$$\partial_{\iota_i}\chi_j = \partial_{\iota_j}\chi_i; \; \partial_{\chi_i}\chi_j = \partial_{\iota_j}\iota_i; \; \partial_{\chi_i}\iota_j = \partial_{\chi_j}\iota_i \qquad i, j = 1, 2, \ldots \tag{4.1.34}$$

In the linear range, the Onsager symmetry guarantees the existence of the entropy production as a scalar thermodynamic potential. The first and third sets of integrability conditions are just the Onsager relations that are observed to be the non-equilibrium analogues of the equilibrium Maxwell relations. These relations may hold outside the linear range and when they do we have a scalar potential, otherwise not.

The second set of integrability conditions are obviously false. They were obtained by taking the ith and jth term from the first and second sums in (4.1.33). From this we deduce that *the entropy production cannot be a function of both the intensive force variables and the extensive velocity variables*. Moreover, the entropy production cannot be a function of all the intensive variables alone, since the size of the system cannot be expressed solely in terms of the intensive variables. Consequently, we must put

$$\iota_j d\chi_j = 0 \tag{4.1.35}$$

which bears a striking resemblance to the equilibrium Gibbs – Duhem relation (1.3.7) We recall from section 1.3 that the Gibbs – Duhem relation provides a relationship among the intensive variables in differential form. In fact, equation (4.1.35) is the non-equilibrium counterpart of the equilibrium Gibbs – Duhem relation (Lavenda, 1972). From equilibrium thermodynamics, we know that the N equations of state

$$\chi_i = \chi_i(\alpha_1, \ldots, \alpha_N), \qquad i = 1, \ldots, N \tag{4.1.36}$$

permit us to obtain the remaining equation of state by integrating the Gibbs – Duhem equation which introduces an undetermined constant of integration. Likewise, a knowledge of the N phenomenological relations

$$\chi_i = \chi_i(\iota_1, \ldots, \iota_N), \qquad i = 1, \ldots, N \tag{4.1.37}$$

allow us to obtain the $N+1$th relation by integration of (4.1.35).

Dividing by the scale factor ι_{N+1} in equation (4.1.35), we obtain

$$\mathrm{d}\chi_{N+1} = -\iota'_j \mathrm{d}\chi_j \tag{4.1.38}$$

where $\iota'_j = \iota_j/\iota_{N+1}$. Equations (4.1.37) can be solved for the velocities

$$\iota'_j = \iota'_j(\chi_1, \ldots \chi_N), \qquad j = 1, \ldots, N \tag{4.1.39}$$

provided that the Jacobian

$$J = \frac{\partial(\chi_1, \ldots, \chi_N)}{\partial(\iota'_1, \ldots, \iota'_N)} \neq 0 \tag{4.1.40}$$

On account of the non-equilibrium Gibbs – Duhem relation (4.1.35), we know that the entropy production is a first order homogeneous function of the velocities

$$\sigma_S = \partial_{\iota_j}\sigma_S \iota_j = -\chi_j \iota_j \tag{4.1.41}$$

where the sum runs over the $N+1$ components. Introducing the phenomenological relations in the form (4.1.39) into (4.1.41), we obtain χ_{N+1} as a function of all the other N forces

$$-\chi_{N+1} = \zeta(\chi_1, \ldots, \chi_N) \tag{4.1.42}$$

where it is convenient to introduce a minus sign.

The function $\zeta(\chi)$ is an integral of equation (4.1.38) which implies that the integrability conditions

$$\partial_{\chi_i}\iota'_j = \partial_{\chi_j}\iota'_i \qquad i, j = 1, 2 \ldots \tag{4.1.43}$$

are satisfied. We recognise (4.1.43) as the third set of integrability conditions (4.1.34). Moreover, the function $\zeta(\chi)$ is just the Legendre transform of the entropy production with respect to the N velocities

$$\zeta(\chi) = \partial_{\iota'_j}\sigma_S \iota'_j - \sigma_S(\iota') \tag{4.1.44}$$

Its functional dependence is displayed by taking the differential of (4.1.44)

$$\mathrm{d}\zeta(\chi) = -\iota'_j \mathrm{d}\chi_j \tag{4.1.45}$$

which is none other than the Glansdorff – Prigogine $\mathrm{d}_\chi \sigma_S$ function.

Since we have lost no information in performing a Legendre transform, why is the function $d\zeta(\chi)$ preferred to the differential of the entropy production $d\sigma(\iota)$ in the derivation of a universal criterion of evolution? To find the answer, we need not consider a 'universal' criterion but only a 'local' one. We expand both functions about a given non-equilibrium stationary state. Using the symbolism invented by Glansdorff and Prigogine (1954, 1964) we distinguish between large 'd' and small 'δ' variations and write to second order:

$$\delta_X\sigma_S \equiv \delta\zeta(\chi) = -\iota_j'^0\chi_j - \tfrac{1}{2}(\partial_{\chi_i}\iota_j')\chi_i\chi_j \tag{4.1.46}$$

$$\delta_\iota\sigma_S \equiv \delta\sigma_S = -\chi_j^0\iota_j - \tfrac{1}{2}(\partial_{\iota_i}\chi_j)\,\iota_i\iota_j \tag{4.1.47}$$

Provided things can be arranged so that the velocities vanish at the stationary state, the expansion for $\delta\zeta(\chi)$ about a non-equilibrium stationary state begins quadratic terms, as opposed to $\delta\sigma_S$ which starts off with a linear term. The dynamic form of Le Châtelier's principle requires (cf. section 6.1)

$$(\partial_{\chi_j}\iota_j')_{\chi_i \neq \chi_j} \leq 0 \tag{4.1.48}$$

which, expressed in words, says that the velocities are a decreasing function of their conjugate forces. Consequently, the local form of the Glansdorff–Prigogine (1954) universal criterion of evolution is

$$\delta_\chi\sigma_S \equiv \delta\zeta(\chi) \geq 0 \tag{4.1.49}$$

and this is the reason why the function $d\zeta(\chi)$ is preferred to the differential of the entropy production. We have shown it to be a simple Legendre transform of the entropy production.

4.2 Thermodynamic *versus* kinetic stability criteria

In the more recent formulations of the generalised thermodynamic stability criteria, attention has been turned to the local forms of the thermodynamic stability criteria (Glansdorff, Nicolis and Prigogine, 1974). This formulation permits a direct comparison with linear stability analysis in terms of the first and second methods of Liapounov (Lavenda, 1970). Indeed, we would hope that the thermodynamic stability criteria would be more general than the kinetic stability criteria, since they have been derived, in some way or another, from the second law. Unfortunately, the generalised thermodynamic criteria do not provide sufficient conditions for stability or instability. In more general terms, any thermodynamic theory that derives its stability criteria solely in terms of scalar potentials is doomed to failure (section 7.4). There are many types of non-equilibrium process that cannot be described solely in terms of scalar potentials (cf. chapter 8). However, let us consider the thermodynamic and kinetic merits of the Glansdorff–Prigogine criteria.

The rate of change of the specific entropy is expanded about a non-equilibrium stationary state and, provided the stationary state values of the velocities vanish, we obtain

$$\partial_t\eta = \partial_t\delta\eta + \tfrac{1}{2}\partial_t\delta^2\eta + (\text{higher order terms}) \tag{4.2.1}$$

where

$$\partial_t \delta\eta = -\chi_j^0 \iota_j \tag{4.2.2}$$

$$\tfrac{1}{2}\partial_t \delta^2\eta = -\chi_j \iota_j \tag{4.2.3}$$

Observe that (4.2.1) is just the Glansdorff–Prigogine $\delta_\iota \sigma_S$ function. On account of the linear term, this term could not be given a well-determined sign as a thermodynamic criterion of stability. The claim is that the positive semidefiniteness of (4.2.3) is the fundamental criterion of stability in the nonlinear range of thermodynamics (Glansdorff and Prigogine, 1971). This means that *both* the stability criteria of linear and nonlinear thermodynamics is pictorially determined by the curvature of the surface of the rate of change of the entropy in the generalised Gibbs space.

The thermodynamic criteria of stability and instability have been succinctly summarised (Glansdorff, Nicolis and Prigogine, 1974) as

$$\tfrac{1}{2}\partial_t \delta^2 \eta > 0 \quad \Rightarrow \quad \text{asymptotic stability} \tag{4.2.4}$$

$$\tfrac{1}{2}\partial_t \delta^2 \eta < 0 \quad \Rightarrow \quad \text{instability} \tag{4.2.5}$$

$$\tfrac{1}{2}\partial_t \delta^2 \eta = 0 \quad \Rightarrow \quad \text{marginal stability} \tag{4.2.6}$$

where a 'marginally' stable state is one with indeterminate stability characteristics; it separates stable from unstable regimes. The condition attached to (4.2.4–6) is that they must be evaluated 'along the motion'. What this is intended to mean is that the 'excess' entropy production (which is identical to the rate of change of the second variation of the entropy) has to be evaluated using a specific set of kinetic equations of the form (4.1.9). In this manner, kinetics is supposed to be introduced into an otherwise purely thermodynamic criterion. Writing (4.2.1) in the form

$$\partial_t \eta = \overset{0}{\chi_j} \iota_j + \tfrac{1}{2}\Lambda_{ij}\iota_i\iota_j \tag{4.2.7}$$

the generalised thermodynamic stability criterion (4.2.4) states that the surface of the entropy production in the generalised Gibbs space must be concave. The generalised Gibbs space is spaced by all the independent velocity variables.

For any fixed set of values of the velocities in (4.2.7), $\partial_t \eta = \text{const}$ and (4.2.7) may define an N-dimensional ellipsoid in the generalised Gibbs space. If we rotate the coordinate system so that each new axis coincides with a principal line of the quadratic form, then (4.2.7) takes the canonical form

$$\partial_t \eta = -x_j \iota'_j + \tfrac{1}{2}\lambda_j \iota_j'^2 \tag{4.2.8}$$

Considering the case that the discriminant of the quadratic form $\Delta = \lambda_1 \lambda_2 \ldots \lambda_N \neq 0$, then by a translation

$$\iota'_j = \left(\iota''_j + \frac{x_j}{\lambda_j}\right) \tag{4.2.9}$$

we obtain

$$\partial_t \eta = -\tfrac{1}{2}\frac{x_j^2}{\lambda_j} + \tfrac{1}{2}\lambda_j \iota_j''^2 \tag{4.2.10}$$

The excess entropy production will be positive provided all the characteristic numbers λ_j are positive. We therefore conclude that the curvature of the surface of the thermodynamic potential in the generalised Gibbs space is still determined by their second variations. Moreover, this provides a pictorial interpretation of the stability criteria (4.2.4–6): in generalised thermodynamics, instability implies the inversion of the surface of entropy production in the generalised Gibbs space.

However plausible the generalised thermodynamic stability criteria may be, (4.2.10) shows that it cannot be justified from the second law. Justification of the generalised thermodynamic criteria must therefore be sought it terms of a comparison with linear kinetic stability analysis (Lavenda, 1970; Keizer and Fox, 1974; Glansdorff, Nicolis and Prigogine, 1974). On a number of occasions (Nicolis, 1971; Glansdorff, Nicolis and Prigogine, 1974; Prigogine and Lefever, 1975) the analogy was drawn between the excess entropy and Liapounov function. The 'direct' method of Liapounov consists of a search for a function V which is definite in a certain domain, including the stationary state, while the time derivative $\dot{V}$ is required to be semidefinite and have an opposite sign (Minorsky, 1962) (cf. section 7.1). Consequently, there is nothing unique about a Liapounov function. In generalised thermodynamics V is to be associated with the excess entropy (4.1) and $\dot{V}$ with the excess entropy production (4.2.3). The concept of a Liapounov function is much more general than its intended use in generalised thermodynamics. It can provide sufficient conditions of stability or instability over finite regions of configuration space (i.e. the space of the independent variables) and not just in the neighbourhood of the stationary state.

In fact, the entire analogy with a Liapounov function is inappropriate and extremely misleading. Generalised thermodynamics offers inequality (4.1) as a *postulate* based on the assumption of local equilibrium. It has nothing to do with, and has never been shown to be valid for, a particular set of kinetic equations. The entire criterion of stability lies in inequality (4.2) and hence the analogy with a Liapounov function is unjustified. Furthermore, the generalised thermodynamic stability criteria (4.2.4–6) have the appearance of complete generality. A little closer inspection shows that (4.2) will provide a sufficient condition for stability or instability when it 'is evaluated at the critical point by eliminating all but one variable, using the (linearised) kinetic equations for a single normal mode' (Glansdorff, Nicolis and Prigogine, 1974). A simple illustration of the thermodynamic criterion will show that even in this case we do not obtain sufficient conditions for stability or instability.

The thermodynamic expression (4.2.3) for the excess entropy production has to be translated into kinetic terms. The thermodynamic force is defined by (4.1.3) or equivalently

$$-\delta\eta = \chi_j\xi_j = R(\log \alpha_j)\xi_j \tag{4.2.11}$$

where ξ_j is the perturbation from a time-independent state

$$\xi_j = \alpha_j - \alpha_j^0 \tag{4.2.12}$$

The stationary value of the internal state variable is denoted by α_j^0 and R is the universal gas constant. Writing $X_j = \chi_j^0 + \chi_j$ we obtain from (4.2.11) the expression of the variation in the thermodynamic force

$$\chi_j = R\xi_j/\alpha_j^0 \tag{4.2.13}$$

Linearising the kinetic equations (4.1.9) about a given non-equilibrium stationary state, we obtain

$$\iota_i = \beta_{ij}\xi_j \tag{4.2.14}$$

where the differential coefficients $\beta_{ij} = (\partial f_i/\partial \alpha_j)_0$ are evaluated at the stationary state. Substituting (4.2.13) and (4.2.14) into the expression for the excess entropy production (4.2.3) leads to

$$\tfrac{1}{2}\partial_t \delta^2 \eta = -\frac{R}{\alpha_j^0}\beta_{ij}\xi_i\xi_j \tag{4.2.15}$$

For our purposes, it will suffice to treat the case of two internal state variables. Using the conventional linear stability analysis (cf. section 7.1), we obtain the necessary and sufficient conditions of stability

$$(\beta_{11} + \beta_{22}) < 0 \tag{4.2.16}$$

$$(\beta_{11}\beta_{22} - \beta_{12}\beta_{21}) > 0 \tag{4.2.17}$$

The quadratic form (4.2.15) can be analysed in terms of the classic theory of second order phase transitions (Tisza, 1966). It can be transformed into the canonical form

$$\tfrac{1}{2}\partial_t \delta^2 \eta = -\frac{R}{\alpha_j^0}\tau_j \xi_j'^2 \tag{4.2.18}$$

which shows that the quadratic form will be positive definite if, and only if,

$$\tau_j < 0, \qquad j = 1, 2, \ldots \tag{4.2.19}$$

At a 'critical' point, one of the τ_j coefficients vanish.

The τ_j coefficients can be expressed in terms of the principal minors of the matrix $[\beta_{ij}]$

$$D_2 = \begin{vmatrix} \beta_{11} & \beta_{12} \\ \beta_{21} & \beta_{22} \end{vmatrix} \tag{4.2.20}$$

The conditions that the quadratic form (4.2.15) will be positive definite are

$$\tau_1 = \beta_{11} < 0 \tag{4.2.21}$$

$$\tau_2 = \frac{D_2}{D_1} = \frac{\beta_{11}\beta_{22} - \beta_{12}\beta_{21}}{\beta_{11}} < 0 \tag{4.2.22}$$

The second condition is identical to (4.2.17) provided the first condition is satisfied.

At first glance we would conclude that stability is 'threatened' or 'compromised' (to use the terminology of generalised thermodynamics) when parameters on which β_{11} depend cause it to change sign. However, Glansdorff, Nicolis and Prigogine (1974) claim that this is definitely the wrong way to handle (4.2.15). We forgot to introduce the kinetics by evaluating the quadratic form with the aid of the linearised kinetic questions (4.2.14). In other words, the perturbations ξ_i and ξ_j cannot be treated as being independent. Although this goes against the entire spirit of thermodynamic analyses of quadratic forms, let us see what these authors mean.

The first thing to point out is that we cannot treat a general form of the perturbation; rather we are confined to represent the perturbation as a normal mode with zero frequency. The reason for this is obvious: if we considered a normal mode of arbitrary frequency, this would introduce an undetermined parameter into the quadratic form and we could not obtain a condition of stability or instability without having first to solve the secular equation. But if we resolve the secular equation then we have no need whatever of the thermodynmic criterion.

The prescription then is to introduce a normal mode of zero frequency into equations (4.2.14). This has the effect of interrelating the perturbation amplitudes. For $N = 2$, we obtain

$$\xi_2 = -\frac{\beta_{21}}{\beta_{22}}\xi_1 \quad \text{and} \quad \xi_1 = -\frac{\beta_{12}}{\beta_{11}}\xi_2 \tag{4.2.23}$$

Here, we must be careful: only one of the two conditions (4.2.23) can be introduced into the quadratic form

$$\tfrac{1}{2}\partial_t\delta^2\eta = -R\left\{\frac{1}{\alpha_1^0}[\beta_{11}\xi_1 + \beta_{12}\xi_2]\xi_1 + \frac{1}{\alpha_2^0}[\beta_{21}\xi_1 + \beta_{22}\xi_2]\xi_2\right\} \tag{4.2.24}$$

If we introduce the first condition (4.2.23) into (4.2.24) we obtain

$$\tfrac{1}{2}\partial_t\delta^2\eta = -\frac{R}{\alpha_1^0}\left[\beta_{11} - \frac{\beta_{12}\beta_{21}}{\beta_{22}}\right]\xi_1{}^2 \geq 0 \tag{4.2.25}$$

while if we introduce the second condition we get

$$\tfrac{1}{2}\partial_t\delta^2\eta = -\frac{R}{\alpha_2^0}\left[\beta_{22} - \frac{\beta_{21}\beta_{12}}{\beta_{11}}\right]\xi_2{}^2 \geq 0 \tag{4.2.26}$$

Observe that the second condition of stability is precisely (4.2.22) that we have obtained by the thermodynamic method which treats the perturbations as if they were independent! Although we have used the method of Glansdorff and Prigogine (1971),we again find the Keizer–Fox (1974) result that the limits of stability, as given by the excess entropy production, not only depend upon the determinant of the secular equation but also on the parameters for which β_{11} changes sign. This observation had been noticed much earlier (Lavenda, 1970). In neither of the two analyses of the quadratic form of the excess entropy production do we obtain the necessary and sufficient criteria of stability as determined from kinetic analysis, (4.2.16) and (4.2.17). We can therefore conclude that *even* in the particular case of a zero frequency normal mode, the thermodynamic criterion of the excess entropy production is meaningless.

Generalised thermodynamics also claims to be able to treat the more general case of a non-zero frequency normal mode (Glansdorff and Prigogine, 1971; Nicolis, 1971). It is observed that for a single mode of complex frequency, the amplitudes of the perturbations are also complex. The analysis is handled in terms of the Glansdorff–Prigogine universal criterion of evolution (4.1.26). Now, in order that ${}_X\dot{\sigma}_S$ turn out to be real, we have to write (4.1.26) in the form

$${}_X\dot{\sigma}_S = -\tfrac{1}{2}[\iota^*\cdot\dot{\chi} + \iota\cdot\dot{\chi}^*] \tag{4.2.27}$$

where the asterisk denotes the complex conjugate. For a single normal mode both ι_j and χ_j are taken to be proportional to

$$\exp(\omega_1 + i\omega_2)t \tag{4.2.28}$$

so that the expression for the universal criterion of evolution (4.2.27) becomes

$$_{X}\dot{\sigma}_S = -\tfrac{1}{2}\omega_1(\boldsymbol{\iota}^*\cdot\boldsymbol{\chi} - \boldsymbol{\iota}\cdot\boldsymbol{\chi}^*) - \frac{i\omega_2}{2}(\boldsymbol{\iota}^*\cdot\boldsymbol{\chi} - \boldsymbol{\iota}\cdot\boldsymbol{\chi}^*) \leqq 0 \tag{4.2.29}$$

Observe that the first term is ω_1 times the excess entropy production

$$\delta^2\sigma_S = -\tfrac{1}{2}(\boldsymbol{\iota}^*\cdot\boldsymbol{\chi} + \boldsymbol{\iota}\cdot\boldsymbol{\chi}^*) \tag{4.2.30}$$

which is also equal to $\frac{1}{2}\partial_t\delta^2\eta$. The term which is multiplied by ω_2 has been called the 'mixed entropy production' (Glansdorff and Prigogine, 1971)

$$\sigma_S{}^M = -\tfrac{i}{2}(\boldsymbol{\iota}^*\cdot\boldsymbol{\chi} - \boldsymbol{\iota}\cdot\boldsymbol{\chi}^*) \tag{4.2.31}$$

The thermodynamic stability criteria for a single mode are

$$\omega_1\delta^2\sigma_S = \omega_1\tfrac{1}{2}\partial_t\delta^2\eta = \omega_1{}^2\delta^2\eta \leqq 0 \tag{4.2.32}$$

and the direction of rotation is fixed by

$$\omega_2\sigma_S{}^M = {}_{X}\dot{\sigma}_S - \omega_1\delta^2\sigma_S$$

$$= (\omega_1{}^2 + \omega_2{}^2)\delta^2\eta - \omega_1{}^2\delta^2\eta = \omega_2{}^2\delta^2\eta \leq 0 \tag{4.2.33}$$

How a scalar determines the direction of rotation is not made clear; it is supposedly intended as a new thermodynamic principle. Let us apply the criterion to a simple example.

Suppose that the linearised kinetic equations are

$$\dot{\xi}_1 = -\beta_{12}\xi_2 \tag{4.3.34}$$

$$\dot{\xi}_2 = \beta_{12}\xi_1 \tag{4.2.35}$$

and, for simplicity, we put $\alpha_1^0 = \alpha_2^0 = \alpha^0$. The variation in the thermodynamic force is given by (4.2.13). From linear stability analysis, we calculate the frequency of oscillation to be $\omega_2^{\pm} = \pm\beta_{12}$; the frequency of damping vanishes in this state. The mixed entropy production (4.2.31) then becomes

$$\sigma_S{}^M = -\frac{i}{\alpha^0}R\beta_{12}(\xi_1^*\xi_2 - \xi_1\xi_2^*) \tag{4.2.36}$$

According to criterion (4.2.33), we multiply the mixed entropy production by either one of the frequencies and require the product to be less than zero. This 'fixes the direction of rotation around the steady state' (Nicolis, 1971). If this criterion has any meaning at all, we have a fifty-fifty chance of coming out with the correct answer.

The thermodynamic field analyses of oscillatory processes are given in chapter 8, where it will become apparent that thermodynamic analyses performed solely in terms of scalar potentials are incapable of analysing the physical processes. The important point to bear in mind is that there are strict limitations on the use of scalar thermodynamic potentials and it is incorrect to use them beyond their range of validity. Although generalised thermodynamics has failed to produce a valid set of thermodynamic stability criteria, it has clearly emphasised the need of a sound and self-consistent formulation of non-equilibrium thermodynamics that would be capable of characterising non-equilibrium processes occurring in the nonlinear range of thermodynamics. In the next chapter will be considered how such a thermodynamic theory can be formulated, accepting the view of rational thermodynamics that any macroscopic theory of non-equilibrium thermodynamics, when properly formulated, should delineate its range of validity.

References

Glansdorff, P. and Prigogine, I. (1954). Sur les propriétés différentielles de la production d'entropie. *Physica,* **20,** 773–780.

Glansdorff, P. and Prigogine, I. (1964). On a general evolution criterion in macroscopic physics. *Physica,* **30,** 351–374.

Glansdorff, P. and Prigogine, I. (1971). *Thermodynamic Theory of Structure, Stability, and Fluctuations.* Wiley-Interscience, New York.

Glansdorff, P., Nicolis, G. and Prigogine, I. (1974). The thermodynamic stability theory of non-equilibrium states. *Proc. natn. Acad. Sci. U.S.A.,* **71,** 197–199.

de Groot, S. and Mazur, P. (1962). *Non-equilibrium Thermodynamics.* North-Holland, Amsterdam.

Keizer, J. and Fox, R. F. (1974). Qualms regarding the range of validity of the Glansdorff – Prigogine criterion for stability of non-equilibrium states. *Proc. natn. Acad. Sci. U.S.A.,* **71,** 192–196.

Landsberg, P. T. (1972). The fourth law of thermodynamics. *Nature,* **238,** 229–231.

Lavenda, B. H. (1970). Ph. D. Thesis, University of Bruxelles.

Lavenda, B. H. (1972). Generalized thermodynamic potentials and universal criteria of evolution. *Lett. Nuovo Cimento,* **3,** 385–390.

Mel, H. C. (1954). *Bull. Acad. Roy. Belg. Cl. Sci.,* **40,** 834.

Minorsky, N. (1962). *Nonlinear Oscillations.* Van Nostrand, Princeton.

Nicolis, G. (1971). Stability and dissipative structures in open system far from equilibrium. *Adv. chem. Phys.,* **19,** 209–324.

Prigogine, I. (1947). *Etude Thermodynamique des Phénomènes Irréversibles.* Editions Desoer, Liège.

Prigogine, I. (1949). Le domaine de validité de la thermodynamique des phénomènes irréversibles. *Physica,* **15,** 272–284.

Prigogine, I. (1954). *Bull. Acad. Roy. Belg. Cl. Sci.,* **31,** 600.

Prigogine, I. (1967). *Introduction to Thermodynamics of Irreversible Processess,* 3 ed. Wiley–Interscience, New York.

Prigogine, I and Lefever, R. (1975). Stability and self-organization in open systems. *Adv. chem. Phys.,* **29,** 1–28.

Tisza, L. (1966). *Generalized Thermodynamics.* MIT Press, Cambridge, Mass.

5 Nonlinear Thermodynamics

In the preceding two chapters there were discussed two formulations of non-equilibrium thermodynamics that are based on inequalities resulting from the second law. At this point, it is well worth while to discuss what can be accomplished. The prescription for deriving the explicit expression for the second law inequality is by now standard and can be found in most textbooks (for example, de Groot and Mazur, 1962) (cf. section 9.2). The second law may, under certain circumstances, be able to tell us which processes can occur but it certainly does not describe how they evolve or provide the necessary and sufficient conditions for the stability of nonlinear thermodynamic processes. It is therefore necessary to determine the principles governing thermodynamic processes occurring in the nonlinear range.

This chapter presents an approach to nonlinear thermodynamics that is based on the thermodynamic principle of the balance of power. Whereas in linear thermodynamics the system is isolated and the entropy production will always appear as the energy which is dissipated, we now have external forces that act on the system to prevent it relaxing to thermodynamic equilibrium. We can therefore envisage that the nonlinear thermodynamic principle will describe a balance of power flows between the system and its environment.

Once the nonlinear thermodynamic principle is known, the stability of the thermodynamic process can be determined. In contrast to rational thermodynamics we make a clear distinction between linear and nonlinear thermodynamics. Unlike generalised thermodynamics, the approach based on the thermodynamic principle of the balance of power is entirely macroscopic in nature. It does not look for justification in terms of microscopic arguments that cannot always be applied to macroscopic processes.

Our approach to nonlinear thermodynamics is based on the following two assumptions:

(1) Primitive thermodynamic quantities, such as temperature, entropy or internal energy, are not in need of definition or existence of proof.

(2) All thermodynamic potentials can be calculated from a sufficient number of static measurements made on the system.

The first assumption precludes a rather sterile discussion of whether thermodynamic potentials can be defined out of equilibrium (Meixner, 1941, 1943). If we consider the statistical mechanical approach which correlates the entropy with the H-function of Boltzmann, it is easy to obtain a meaning of the entropy that is not impaired by conditions of equilibrium. Our feeling is that there is no conceptual difficulty in defining the entropy of a non-equilibrium state, although there may arise some problems as to its measurement.

Assumption (1) implies that we intend to develop nonlinear thermodynamics from a Gibbsian point of view (Lavenda, 1972, 1974), rather than the axiomatic approach of Carathéodory. The main reason for this choice is that Carathéodory's axiomatic approach does not provide the criteria for its validity. We have already discussed this point at length in section 1.1.

The second assumption preserves the equilibrium character of the thermodynamic potentials. In particular, it excludes the possibility that the internal energy or the entropy is functionally dependent on the rate variables (that is, the time derivatives of the extensive thermodynamic variables). If this were not the case, then we would be left with only a formal resemblance between equilibrium thermodynamic potentials and their non-equilibrium extensions. Let it be clear this does not imply that we are limiting ourselves to a type of 'local' equilibrium assumption, as is done in generalised thermodynamics. We do suppose that a certain type of Gibbs relation is valid in nonlinear thermodynamics. Furthermore we state honestly and bluntly that such a relation will serve only one purpose: to eliminate conservative terms in the equation of the balance of power. If we could devise other means for separating conservative and dissipative terms, they would serve equally as well provided there is no destruction of the equilibrium functional dependencies of the thermodynamic potentials.

The criteria that we use to determine the form of the 'generalised' Gibbs equation are (a) in the internal energy representation, the energy is a function of *all* the variables that are associated with the performance of work, and (b) the internal energy is not a function of the rate variables. Hence, we introduce 'pseudo-thermodynamic' variables (Tisza, 1966) into the Gibbs equation whose variation describes the performance of work due to all the conservative processes. The pseudo-thermodynamic variables are associated with external fields or stresses that act on the system and, by their very nature, vanish in the state of equilibrium. Consequently, they do not destroy the continuity between equilibrium and non-equilibrium thermodynamics. They do, however, contribute to the internal energy of a non-equilibrium system so that it would be incorrect to exclude them in a generalised Gibbs relation. The variations of the pseudo-thermodynamic variables cannot be formalised as exchanges of conserved quantities. Rather, they are subject to boundary and/or initial conditions. In other words, they do not conform to the equilibrium constraint of additive invariance, discussed in section 1.2.

The Gibbs equation is a thermodynamic expression of the principle of the conservation of energy. In this chapter we ask for the balance of power equation that applies to kinetic processes in nonlinear thermodynamics. It cannot be obtained simply by differentiating the generalised Gibbs relation in time, since this equation contains only information concerning the conservative system processes. A key point is how the concept of dissipation is introduced and the generalised Gibbs relation proves to be a useful tool in the separation of conservative and dissipative processes.

In the derivation of the thermodynamic principle of the balance of power, we intend to carry over as much of the equilibrium structure of thermodynamics that is permitted. It is for this reason that we have taken time to discuss the classic formulations of equilibrium thermodynamics in chapter 1. In contrast to rational thermodynamics, since we assume the validity of a conservative constitutive relation, we are now free to use the derived relation—the principle of the balance of power—as a criterion of stability. Moreover, in contrast to generalised thermodynamics, the adaptation of the constitutive relation does not presuppose a stability criterion such as the condition that cellular or local equilibrium should be stable.

In the context in which the generalised Gibbs equation is introduced, it is easy to see why the rate variables have to be excluded. Let us assume the contrary. There then exists a Gibbs—Duhem relation which implies a functional dependence of the intensive thermodynamic variables, such as the temperature, on the rate variables. While the temperature can depend upon the strain, say, it certainly cannot depend upon the rate of strain. This does not however imply that we are going to exclude the rate variables in our formulation; the generalised Gibbs relation has only one part in our approach and it is not fundamental to it. We shall find that the rate variables enter in the formulation in a natural way when we analyse the forces acting on the system (cf. sections 5.1 and 5.2).

A fundamental property of the thermodynamic potentials, that we are going to make extensive use of, is their conditions for existence or 'exactness' conditions. We have seen in chapter 4 that the inability to delineate the range of validity of thermodynamics is a potential source of confusion. Although it has long been recognised, in other branches of macroscopic physics, that it is impossible to describe rotational processes by means of scalar potentials, this seems to have gone virtually unnoticed in non-equilibrium thermodynamics (Coleman and Truesdell, 1960; Glansdorff and Prigogine, 1971).

The criteria for the existence of scalar thermodynamic potentials allow us to shift the emphasis from what are the origins to what are the effects of the symmetries of the phenomenological coefficients. In this respect, we do not have to worry about whether the phenomenological symmetries have been established on the principle of microscopic reversibility at equilibrium, or whether the antisymmetric part of the tensor has been set equal to zero because it has no observable physical consequences (see the discussion in section 2.3 and Casimir, 1945). In the event that the antisymmetric part of a phenomenological coefficient matrix does have an observable physical consequence (for example, rotations about a non-equilibrium stationary state) then the approach to be developed in this chapter does not apply. This will become apparent from the quasi-thermodynamic stability analysis of chapter 7. It simply means that we cannot describe processes caused by non-conservative forces in terms of scalar potentials. This is the reason for including the thermodynamic field analyses of chapter 8.

In summary, the criteria for the existence of scalar thermodynamic potentials not only delineate the range of validity of nonlinear thermodynamics but they are also employed in the same spirit as other theories of macroscopic physics. In this chapter, we ask what are the implications rather than what are the origins of the phenomenological symmetries in nonlinear thermodynamics. These are most easily discernible through an analysis of the balance of power equation.

5.1 The thermodynamic principle of the balance of power

In this section the thermodynamic principle of the balance of power in the energy representation is derived. This will display, in the clearest way possible, the connection with the mechanical principle of the conservation of power.

In addition to the two assumptions given in the preceding section, we will need the balance equations for the specific internal energy (3.3.36) and the specific entropy

$$\dot{\eta}+\frac{1}{\rho}\nabla\cdot(\boldsymbol{q}/\theta)=\sigma_S \tag{5.1.1}$$

where, as usual, the dot symbol is used to denote the substantial derivative. For more details see chapter 3. We recognise that these are not the most general forms of the balance equations; for instance, the entropy current consists of the heat flux vector only. In section 9.1 are considered more general forms of the balance equations, including that of momentum, when we consider the thermodynamics of the continua, for example, equation (9.1.16). Here, we choose the simplest exposition, which in rational thermodynamics would be referred to as the analysis of 'simple' materials. As usual, we consider the statement $\sigma_S \geq 0$ as a local formulation of the second law: the entropy generated locally in the system cannot be negative irrespective of whether the system is isolated or not.

When the divergence of the heat flux vector is eliminated between the balance equations for the specific internal energy (3.3.36) and the specific entropy (5.1.1) there results

$$\theta(\pi-\sigma_S)=\dot{\varepsilon}-\theta\dot{\eta}+\frac{1}{\rho\theta}\boldsymbol{q}\cdot\boldsymbol{g}-\frac{1}{\rho}tr\{\mathbf{T}\,\mathbf{L}\} \tag{5.1.2}$$

We have already obtained an equivalent result in our discussion of rational thermodynamics, cf. equation (3.3.37). The main divergence from the philosophy of rational thermodynamics is to evaluate equation (5.1.2) using a generalised Gibbs equation that we assume to be valid from the outset.

The generalised Gibbs relation will take the form of an expression for the substantial derivative of the thermodynamic potential in whose representation we are working. This relation will serve to define the functional dependency of the thermodynamic potential. In the last paragraph we have given the criteria that will permit us to determine the generalised Gibbs relation. In particular, the exclusion of a functional dependency on the rate variables implies that no acceleration terms will appear in the substantial derivative of the thermodynamic potential. The criteria for determining the functional dependencies of the thermodynamic potential follow from the second assumption. The gist of the argument is that the intensive variables have the same functional dependencies that they do at equilibrium, with the important exception that they can also be functions of the pseudo-thermodynamic variables.

As already mentioned, the point of departure from rational thermodynamics lies in our assumption that there exists a valid constitutive relation for the thermodynamic potential. This will permit us to investigate the stability properties of a *given* thermodynamic process, which is defined by the generalised Gibbs relation, using the power equation (5.1.2). However, we soon come across the same problem encountered in rational thermodynamics: the power equation has to be split into thermal and mechanical power equations. This implies that heat conduction and mechanical dissipation are two inherently different forms of dissipation (cf. section 2.1). Since it is our intention to formulate stability criteria in mechanical terms, based on the analysis of the power equation, it will prove convenient to introduce the specific Helmholtz free energy in (5.1.2):

$$\theta(\pi - \sigma_S) = \dot{\psi} + \eta\dot{\theta} + \frac{1}{\rho\theta}\boldsymbol{q}\cdot\boldsymbol{g} - \frac{1}{\rho}tr\{\mathbf{T}\,\mathbf{L}\} \tag{5.1.3}$$

In this way the temperature becomes an independent variable.

The generalised Gibbs equation is now given by the expression for the substantial derivative of the specific free energy. For the thermodynamic processes considered in the power equation (5.1.3), the generalised Gibbs equation is

$$\dot{\psi} = \partial_\theta\psi\dot{\theta} + \partial_{\mathbf{F}}\psi : \dot{\mathbf{F}} + \partial_{\boldsymbol{\alpha}}\psi\cdot\boldsymbol{\imath} \tag{5.1.4}$$

where

$$\partial_\theta\,\psi = \eta \tag{5.1.5}$$

$$\rho\mathbf{F}\partial_{\mathbf{F}}\psi = \mathbf{T}_E \tag{5.1.6}$$

Definition (5.1.5) is analogous to the equilibrium definition, except for the allowance that the specific entropy is to be a function of the strain. The strain is an example of a pseudo-thermodynamic variable. In the more general case, the specific entropy can be a function of all or some of the pseudo-thermodynamic variables. Other examples of pseudo-thermodynamic variables are the electric displacement and magnetic induction vectors.

Definition (5.1.6) states that the specific free energy determines only the elastic part of the stress which is rate independent. This, we recall, is one of the results of rational thermodynamics. Here, we impose this condition from the outset on the basis that the Gibbs–Duhem relation would lead to incorrect functional dependencies of the dependent thermodynamic variables.

Introducing the generalised Gibbs relation (5.1.4) into the power equation (5.1.3) eliminates the non-dissipative terms on the right-hand side of the equation, namely

$$\theta(\pi - \sigma_S) = \left[\partial_{\mathbf{F}}\psi - \frac{1}{\rho}\mathbf{T}(\mathbf{F}^T)^{-1}\right] : \dot{\mathbf{F}} + \partial_{\boldsymbol{\alpha}}\psi\cdot\boldsymbol{\imath} + \frac{1}{\rho\theta}\boldsymbol{q}\cdot\boldsymbol{g} \tag{5.1.7}$$

In fact, this is a general observation: the introduction of the generalised Gibbs equation into the power equation makes explicit the fact that the difference between the entropy production and the power is equal to the system dissipation. The power equation (5.1.7) brings out the very important fact that, in open systems, entropy production is not synonymous to dissipation. In nonlinear thermodynamics it is therefore possible to have a completely reversible process so long as the external power supply compensates exactly the system dissipation. Until now, the terms 'entropy production' and 'dissipation' have been used interchangeably and no notice has been given to the possibility that a nonlinear thermodynamic process can be completely reversible (cf. section 7.2). This observation has also served as the motivation for treating (5.1.7) as a power equation rather than an expression for the second law inequality.

Equation (5.1.7) becomes a mechanical power equation when we impose the condition that the thermodynamic process be isothermal. Stability properties can then be formulated in a particularly simple manner which exploits the analogy with mechanics. Under isothermal conditions, the power equation (5.1.7) reduces to

$$\theta(\pi-\sigma_S)=\left[\partial_{\mathbf{F}}\psi-\frac{1}{\rho}\mathbf{T}(\mathbf{F}^T)^{-1}\right]:\dot{\mathbf{F}}+\partial_{\boldsymbol{a}}\psi\cdot\boldsymbol{\imath} \tag{5.1.8}$$

Since the heat flux vanishes, $\sigma_S=\dot{\eta}$, and in order to bring out the significance of the power equation (5.1.8) we introduce the expression for the substantial derivative of the specific entropy

$$\dot{\eta}=\partial_{\varepsilon}\eta\dot{\varepsilon}+\partial_{\mathbf{F}}\eta:\dot{\mathbf{F}}+\partial_{\boldsymbol{a}}\eta\cdot\boldsymbol{\imath} \tag{5.1.9}$$

into equation (5.1.8) to obtain

$$\theta\pi=\dot{\varepsilon}+\left[\partial_{\mathbf{F}}\varepsilon-\frac{1}{\rho}\mathbf{T}(\mathbf{F}^T)^{-1}\right]:\dot{\mathbf{F}}+\partial_{\boldsymbol{a}}\varepsilon\cdot\boldsymbol{\imath} \tag{5.1.10}$$

Equation (5.1.10) bears a striking resemblance to an energy balance. It states that the rate of change of the energy is equal to the difference between the supplied power and the internal dissipation. If we define a new rate variable $\dot{\boldsymbol{\lambda}}$ which has components $(\dot{\mathbf{F}}, \boldsymbol{\imath})$ then we have, as the definition of the generalised force,

$$\partial_{\dot{\lambda}}\varepsilon=-X \tag{5.1.11}$$

in the energy representation.

Assuming that the velocities are small, the generalised force can be expanded in powers of them

$$X=X(\boldsymbol{a},\mathbf{F})-\boldsymbol{\Lambda}\dot{\boldsymbol{\lambda}}+\ldots \tag{5.1.12}$$

where $\boldsymbol{\Lambda}$ is the phenomenological coefficient matrix in the energy representation. The first term is the force acting for $\dot{\boldsymbol{\lambda}}=\boldsymbol{0}$; that is, for adiabatic motion η = constant. Multiplying (5.1.12) by the velocity vector, the balance of power equation (5.1.10) can be written as

$$\theta\pi=2\theta\phi+\dot{\varepsilon},\qquad \eta=\text{const} \tag{5.1.13}$$

where $\theta\phi$ is the dissipation function in the energy representation

$$\theta\phi=\tfrac{1}{2}\Lambda_{ij}\dot{\lambda}_i\dot{\lambda}_j \tag{5.1.14}$$

Equation (5.1.13) is an expression of the thermodynamic principle of the balance of power in the energy representation. It separates the absorbed power, at constant entropy, into the dissipated energy and the time-rate-of-change of the stored energy, that is, the internal energy. It clearly displays the relation of the thermodynamic principle to the mechanical principle of the conservation of power. On occasion we shall refer to it as the mechanical power balance equation in the internal energy representation. However, it is not well adapted to the analysis of nonlinear thermodynamic processes because of the condition of constant entropy. An easier condition to meet is the constancy of the internal energy. It is for this reason that an analogous expression of the thermodynamic principle of the balance of power in the entropy representation will be derived in the following section.

In accordance with the generalised force expansion (5.1.12) which supposes that the *independent* velocities vanish in a non-equilibrium stationary state, the dissipation function (5.1.14) is seen to be a second order quantity. The power equation (5.1.13) thus contains terms of three different orders of magnitude. The zero order terms

$$\theta\pi^0 = \dot{\varepsilon}^0 \tag{5.1.15}$$

give the condition for the formation of a non-equilibrium stationary state. At least one velocity must be maintained constant by the external constraints in order to prevent the system from relaxing back to equilibrium. The sum in (5.1.14) is over all the independent velocities and does not contain the externally maintained, constant velocity. In other words, the imposed constraints do not contribute to the system dissipation.

The first order term in the mechanical power equation reduces to the stationary state force balance law

$$\boldsymbol{\chi}_E^0 = \boldsymbol{\chi}^0(\boldsymbol{a}^0, \mathbf{F}^0) \tag{5.1.16}$$

while the second order power equation is

$$\theta(\delta^2\pi - 2\phi) = \tfrac{1}{2}\delta^2\dot{\varepsilon} \tag{5.1.17}$$

In order to obtain (5.1.17), we have written the rate independent component of the generalised force term in (5.1.12):

$$\boldsymbol{X}(\boldsymbol{a}, \mathbf{F}) = \boldsymbol{\chi}^0(\boldsymbol{a}^0, \mathbf{F}^0) + \boldsymbol{\chi}(\delta\boldsymbol{a}, \delta\mathbf{F}) \tag{5.1.18}$$

where the first term is the stationary state value of the generalised force and the second term is the variation in the force that is not due to the velocities, namely

$$\tfrac{1}{2}\delta^2\dot{\varepsilon} = -\boldsymbol{\chi}\cdot\boldsymbol{\lambda} \tag{5.1.19}$$

The second order power equation (5.1.17) determines the variational equations and hence the stability properties of the thermodynamic process. However, it will prove more convenient to consider the mechanical power equation in the entropy representation, since irreversibility and dissipation are more clearly discernible.

5.2 The balance equation of mechanical power in the entropy representation

The derivation of the mechanical power equation in the entropy representation begins with equation (5.1.8), which can be written in the more general form as

$$\begin{aligned} \pi - \sigma_S &= \frac{1}{\theta}\{[\partial_\gamma\psi - \boldsymbol{\Gamma}]\cdot\dot{\gamma} + \partial_{\boldsymbol{a}}\psi\cdot\boldsymbol{\imath}\}, \qquad \theta = \text{const} \\ &= \boldsymbol{X}_D\cdot\dot{\boldsymbol{\lambda}} \end{aligned} \tag{5.2.1}$$

where γ represents the set of extensive, pseudo-thermodynamic variables and $\boldsymbol{\Gamma}$ are the conjugate forces. For instance, if γ_i is the strain then Γ_i is the stress, cf. equation (5.1.8). The vector $\dot{\boldsymbol{\lambda}}$ now has the components $(\dot{\gamma}, \boldsymbol{\iota})$, and $\boldsymbol{\lambda}$ can be considered as a generalised displacement vector.

The power equation (5.2.1) is seen to possess two types of force, a thermodynamic force $\boldsymbol{X}_T$, which is a function of state, that is, a function of the λ's and a 'dissipative' force $\boldsymbol{X}_D$ (Lavenda, 1972, 1974), which is a function of their rates of change. In the entropy representation, the thermodynamic force is defined as

$$\partial_{\dot{\lambda}}\sigma_S = (\partial_{\dot{\lambda}}\dot{\eta})_\varepsilon = -\ \boldsymbol{X}_T(\boldsymbol{\lambda}) \tag{5.2.2}$$

which clearly shows that it is defined at constant internal energy. The subscript 'T' will henceforth be used to denote the thermodynamic force. Provided the deviations from a given non-equilibrium stationary state are small, $|\xi/\lambda| \ll 1$, where

$$\xi = \boldsymbol{\lambda} - \boldsymbol{\lambda}^0 \tag{5.2.3}$$

the thermodynamic force can be expanded in terms of these deviations

$$\boldsymbol{X}_T(\boldsymbol{\lambda}) = \boldsymbol{\chi}_T^0 + \boldsymbol{\chi}_T(\xi) \tag{5.2.4}$$

The first term in (5.2.4) is the stationary state value of the thermodynamic force which vanishes at equilibrium. The second term is the variation in the thermodynamic force and, for small deviations from the non-equilibrium, time independent state, it is linear in these deviations

$$\chi_{T_i}(\xi) = S_{ij}\xi_j \tag{5.2.5}$$

where

$$S_{ij} \equiv -\partial_{\lambda_i}\partial_{\lambda_j}\eta \geq 0 \tag{5.2.6}$$

which is the equilibrium stability criterion that the entropy surface in Gibbs space must be convex.

The force on the right-hand side of the power equation (5.2.1) is the dissipative force. We will see that the dissipation function acts as a potential for this force. However, in the special case of constant strain and constant temperature, the dissipative force is the gradient of the specific Helmholtz free energy per unit temperature with respect to the remaining λ variables

$$\frac{1}{\theta}\partial_{\lambda}\psi = \boldsymbol{X}_D, \qquad \dot{\mathbf{F}} = \mathbf{0},\ \dot{\theta} = 0 \tag{5.2.7}$$

This is the closest we can come to defining the dissipative force as the gradient of a classical thermodynamic potential.

Choosing the *independent* velocities so that they vanish at the stationary state, the dissipative force can be expanded in powers of these velocities ($\dot{\boldsymbol{\xi}} = \dot{\boldsymbol{\lambda}}$)

$$\boldsymbol{X}_D(\dot{\boldsymbol{\lambda}}) = \boldsymbol{\chi}_D(\dot{\boldsymbol{\xi}}) + \ldots \tag{5.2.8}$$

where the zero order term in the series expansion vanishes on account of the fact that there is no dissipation for $\dot{\boldsymbol{\lambda}} = \boldsymbol{0}$. The first term is linear in the velocities

$$\boldsymbol{\chi}_{D_i}(\dot{\xi}) = -R_{ij}\dot{\xi}_j \tag{5.2.9}$$

where $R_{ij} = \Lambda_{ij}/\theta$; they are the generalised resistances in the entropy representation.

In order to maintain the system in a non-equilibrium stationary state, there must be at least one velocity that is held constant by the external constraints. This parameter can be identified with the scaling factor in the non-equilibrium Gibbs–Duhem relation (4.1.38). After having derived the power equation in the entropy representation, we shall return to this point in section 5.3. None the less, in view of this fact and the expansion of the thermodynamic force about the stationary state (5.2.4), the entropy production is given by the expansion

$$\begin{aligned}\sigma_S &= \sigma_S^0 + \delta\sigma_S + \delta^2\sigma_S \\ &= -\boldsymbol{\chi}_T^0 \cdot \boldsymbol{\lambda}^0 - \boldsymbol{\chi}_T^0 \cdot \dot{\boldsymbol{\xi}} - \boldsymbol{\chi}_T \cdot \dot{\boldsymbol{\xi}}\end{aligned} \tag{5.2.10}$$

where σ_S^0 is the stationary state value of the entropy production and

$$\delta^2\sigma_S = \tfrac{1}{2}\delta^2\dot{\eta} = -S_{ij}\xi_j\dot{\xi}_i \tag{5.2.11}$$

There is an additional power absorption when the system is displaced slightly from its stationary state. It is accounted for in the expansion of the power function about the stationary state, namely

$$\pi = \pi^0 + \delta\pi + \delta^2\pi \tag{5.2.12}$$

where

$$-\partial_{\boldsymbol{\lambda}}\pi = \boldsymbol{X}_E = \boldsymbol{\chi}_E^0 + \boldsymbol{\chi}_E \tag{5.2.13}$$

The first term in (5.2.12) is the power which the system absorbs in the stationary state; it vanishes when the stationary state happens to be thermodynamic equilibrium. The last term represents the additional power absorbed due to variations in the external forces which are represented by the second term in (5.2.13).

In view of (5.2.8) and (5.2.9), the equation for the balance of mechanical power (5.2.1) can be written as

$$\pi = \sigma_S - 2\phi, \qquad \varepsilon = \text{const} \tag{5.2.14}$$

where ϕ is the dissipation function per unit mass and temperature

$$\phi = \tfrac{1}{2}R_{ij}\dot{\xi}_i\dot{\xi}_j \tag{5.2.15}$$

On account of the fact that the process is isothermal, equation (5.2.14) can also be written as

$$\pi = \dot{\eta} - 2\phi, \qquad \varepsilon = \text{const} \tag{5.2.16}$$

which makes it comparable with the principle of the balance of mechanical power (5.1.13) in the energy representation.

Equation (5.2.16) is a valid expression of the thermodynamic principle of the balance of power in the entropy representation for isothermal, simple materials. As was mentioned in the previous section, it is better adapted to the analysis of nonlinear thermodynamic processes than the corresponding expression (5.1.13) that has been derived in the energy representation. Moreover, it clearly displays the continuity between linear and nonlinear thermodynamics. In fact, it reduces to the linear thermodynamic principle that the dissipated energy is equal to the time-rate-of-change of the entropy for isothermal, simple materials when the external power is shut off, cf. equation (6.2.5).

For isothermal, simple materials the nonlinear thermodynamic principle of the balance of power states that the absorbed power, at constant energy, appears as the difference between the time-rate-of-change of the entropy less that which is dissipated. Expressions (5.2.14) and (5.2.16) are the fundamental results of this chapter. We shall meet them over and over again in the construction of non-equilibrium thermodynamic variational principles and in the stability analysis of nonlinear thermodynamic processes. In the following section some of the properties of the balance equations of mechanical power will be considered.

Of prime importance is the second order balance equation of the mechanical power, since it will determine the stability of a given non-equilibrium stationary state. We appreciate the fact that on account of the expansions of the non-equilibrium thermodynamic potentials about a given non-equilibrium stationary state, (5.2.10) and (5.2.12), the equation of the balance of mechanical power in the entropy representation (5.2.14) will contain terms of three different orders of magnitude. In the preceding section we saw that the same was true for the mechanical power equation that was derived in the energy representation.

Introducing the expansions (5.2.10) and (5.2.12) into equation (5.2.14) and setting equal terms of the same order, we obtain at zero order

$$\pi^0 = \sigma_S^0 \tag{5.2.17}$$

Expressed in words, the stationary value of the absorbed power appears as the entropy produced in the non-equilibrium stationary state. At equilibrium, π^0 vanishes identically and so too does the entropy production.

The first order equation of the balance of mechanical power reduces to the stationary state force balance law, cf. equation (5.1.16) in the energy representation

$$\chi_E^0 = \chi_T^0 \tag{5.2.18}$$

If the stationary state happens to be equilibrium then both (5.2.17) and (5.2.18) vanish identically.

At second order we find

$$\delta^2\pi = \delta^2\sigma_S - 2\phi \tag{5.2.19}$$

which we shall refer to as the 'excess' power equation. It states that the power absorbed when the system is displaced slightly from its stationary state will appear as an excess production of entropy less the energy which is dissipated in the process. The reason that the dissipated energy only appears in the second order power equation is that we have required the independent velocities to vanish in the non-equilibrium stationary state. Is the local thermodynamic principle of the balance of power (5.2.19) also valid for small displacements from the state of equilibrium? The answer depends on what is the cause of the displacement of the system from equilibrium. If the perturbations are applied externally then (5.2.19) is a valid local thermodynamic principle. If the perturbations are due to spontaneous fluctuations in the system then we must set (5.2.19) equal to zero and obtain the local thermodynamic principle of linear thermodynamics which is valid for isolated processes. We shall have a great deal more to say about the local thermodynamic principle (5.2.19) in connection with the quasi-thermodynamic stability criteria to be derived in section 7.4. Let it suffice to remark that the stability of a given non-equilibrium stationary state cannot be determined by the properties of a single non-equilibrium thermodynamic potential. Finally, for isothermal, simple materials the local thermodynamic principle of the balance of power can be expressed as

$$\delta^2\pi = \tfrac{1}{2}\delta^2\dot{\eta} - 2\phi \tag{5.2.20}$$

Inserting (5.2.11) and (5.2.15) into the local thermodynamic principle of the balance of power (5.2.19) and dividing through by the velocities, we obtain the variational equations

$$R_{ij}\dot{\xi}_j + S_{ij}\xi_j = \chi_{Ei}, \qquad i = 1, 2, \ldots, N \tag{5.2.21}$$

for a system with N independent velocities. The variational equations (5.2.21) are the phenomenological equations for a non-equilibrium process that is occurring in an open system. For small deviations from equilibrium that are caused by spontaneous internal fluctuations, the right-hand sides of equations (5.2.21) vanish identically and we obtain the classic phenomenological regression laws of linear thermodynamics (Onsager, 1931) (cf. section 2.2). The variational equations are sufficient to determine the stability properties of the thermodynamic process. This shows that the excess power equation (5.2.19) is inherently connected with the stability properties of the process. Moreover, the variational equations show that dissipation is a first order effect that is accounted for, thermodynamically, in the excess power equation. The next chapter derives the variational equations from a thermodynamic variational principle.

The phenomenological laws of linear thermodynamics ($X_E = 0$) are usually expressed in terms of half-degrees of freedom; that is, first order linear differential equations of the form (5.2.21). However, for many nonlinear thermodynamic processes the damping of the motion is not sufficiently rapid so that the forces are no longer strictly proportional to the velocities. On account of the external constraints, which act as energy reservoirs, or internal destabilising processes, the forces will become proportional to the accelerations, making it necessary to describe the motion in terms of single degree of freedom variational equations (that is second order, linear differential equations).

According to the second assumption, the generalised Gibbs relation precludes a dependency of the specific entropy on the rate variables. In the language of rational thermodynamics, this part of the entropy has been called the 'equilibrium' part. So as no confusion will arise, we shall call this part of the entropy, the 'Gibbsian' part. The 'non-equilibrium' or 'non-Gibbsian' part of the entropy will not enter into a generalised Gibbs relation, since it will be a function of the rate variables. The motivation for introducing a non-Gibbsian part of the entropy is the following: in the entropy representation, the entropy must be a function of all the state variables. If the motion is described in terms of equations of single degrees of freedom, then the initial values of the displacements, as well as the velocities, will be needed to solve the equations. Consequently, the set of state variables, which previously consisted of only the displacements for half-degree of freedom equations, must now be enlarged to include the velocities as well. The specific entropy splits into a Gibbsian part $\eta'(\lambda)$ and a non-Gibbsian part $\eta''(\dot{\lambda})$, cf. (3.4.12) by analogy:

$$\eta(\boldsymbol{\lambda}, \dot{\boldsymbol{\lambda}}) = \eta'(\boldsymbol{\lambda}) + \eta''(\dot{\boldsymbol{\lambda}}) \qquad (5.2.22)$$

Expression (5.2.22) could have been written in the more general form by writing $\eta''(\boldsymbol{\lambda}, \dot{\boldsymbol{\lambda}})$ instead of $\eta''(\dot{\boldsymbol{\lambda}})$. However, such a generalisation will not be required at this time, since we are considering a *linear* formulation of nonlinear thermodynamics for which the phenomenological coefficients are constants.

The substantial derivative of (5.2.22) is

$$\dot{\eta} = \partial_{\boldsymbol{\lambda}}\eta' \cdot \dot{\boldsymbol{\lambda}} + \partial_{\dot{\boldsymbol{\lambda}}}\eta'' \cdot \ddot{\boldsymbol{\lambda}}, \qquad \varepsilon = \text{const} \qquad (5.2.23)$$

At first glance, we would hastily conclude that the thermodynamic force also splits into two parts $\boldsymbol{X}'_T(\boldsymbol{\lambda})$ and $\boldsymbol{X}''_T(\dot{\boldsymbol{\lambda}})$, where

$$-(\partial_{\lambda_i}\eta')_\varepsilon = X'_{Ti}(\boldsymbol{\lambda}) = \chi^0_{Ti} + S_{ij}\xi_j \qquad (5.2.24)$$

$$-(\partial_{\dot{\lambda}_i}\eta'')_\varepsilon = X''_{Ti}(\dot{\boldsymbol{\lambda}}) = M_{ij}\dot{\xi}_j \qquad (5.2.25)$$

Then, in view of the similarity between the dissipative force variation (5.2.9) and (5.2.25), we would arrive at the conclusion that the second member in (5.2.22) is the 'dissipative' entropy (Eringen, 1965) (cf. section 3.4). However, we neglected to use the property that the entropy production is a bilinear sum of forces and conjugate velocities. Hence, the true definition of the thermodynamic force is

$$\begin{aligned} -\boldsymbol{X}_T &= (\partial_{\dot{\boldsymbol{\lambda}}}\dot{\eta})_\varepsilon \\ &= (\partial_{\boldsymbol{\lambda}}\eta')_\varepsilon + \partial_t(\partial_{\dot{\boldsymbol{\lambda}}}\eta'')_\varepsilon \end{aligned} \qquad (5.2.26)$$

It will then be appreciated that when (5.2.26) is multiplied by the velocity vector, it is identical to (5.2.23). Thus,

$$\boldsymbol{X}_T(\boldsymbol{\lambda}, \ddot{\boldsymbol{\lambda}}) = \boldsymbol{X}'_T(\boldsymbol{\lambda}) + \boldsymbol{X}''_T(\ddot{\boldsymbol{\lambda}}) \qquad (5.2.27)$$

where

$$X''_T(\ddot{\boldsymbol{\lambda}}) = -\partial_t(\partial_{\dot{\boldsymbol{\lambda}}}\eta'')_\varepsilon = M_{ij}\ddot{\xi}_j \qquad (5.2.28)$$

and not (5.2.25). The second component of the thermodynamic force (5.2.28) is analogous to a d'Alembertian expression for the forces of inertia in classical mechanics (Machlup and Onsager, 1953). Consequently, $\boldsymbol{X}''(\ddot{\boldsymbol{\lambda}})$ is not dissipative in nature and it is wrong to conclude that $\eta''(\dot{\boldsymbol{\lambda}})$ is the 'dissipative' entropy. In fact, it has the opposite effect.

Inserting (5.2.24) and (5.2.25) into (5.2.23) or, more correctly, multiplying (5.2.26) by the velocity vector, leads to the same result, namely

$$\dot{\eta} = \dot{\eta}^0 - \chi^0_{Ti}\dot{\xi}_i - (S_{ij}\xi_j + M_{ij}\ddot{\xi}_j)\dot{\xi}_i \qquad (5.2.29)$$

We recall that the stationary state contribution to the entropy production is due to the fact that at least one of the components of the velocity vector must be maintained constant in order to prevent the system from relaxing back to equilibrium. Integrating (5.2.29) in time, we obtain, in particular, the expression for the excess entropy

$$\delta^2\eta = -(S_{ij}\xi_i\xi_j + M_{ij}\dot{\xi}_i\dot{\xi}_j) \qquad (5.2.30)$$

which shows that it is a quadratic function of both the generalised displacements and velocities. Finally, introducing the time derivative of expression (5.2.30) into the excess power equation (5.2.20) leads to

$$M_{ij}\ddot{\xi}_j + R_{ij}\dot{\xi}_j + S_{ij}\xi_j = \chi_{Ei}, \qquad i = 1, 2, \ldots, N \tag{5.2.31}$$

which are the single degree of freedom variational equations that replace the half-degree of freedom equations (5.2.21) in nonlinear thermodynamics. In particular, the variational equations (5.2.31) will be used in conjunction with the discussion of the stability and symmetry properties of nonlinear thermodynamic processes in sections 7.4 and 7.5, respectively.

5.3 Properties and forms of the balance equation of mechanical power

This section compares the energy and entropy representations of the mechanical power equations (5.1.13) and (5.2.14), respectively, in regard to the non-equilibrium thermodynamic principles that can be derived from them. Also presented are some interesting alternative forms of the mechanical power equation in the entropy representation that will allow us to clear up a point concerning the expansion of the entropy production (5.2.10) about a given non-equilibrium stationary state.

We want to calculate the maximum work done by the system as it makes a transition from one non-equilibrium state to another. The difference between the mechanical power equation in the energy representation and the balance equation for the energy is that the former allows us to speak of a 'maximum' work done by the system. In the energy balance equation the work done by the system is completely defined; that is, in the absence of a system stress, the work would be equal to the change in the energy of the system. On the other hand, the power equation introduces the notion of dissipation and it then becomes a meaningful question to ask what is the maximum work that can be performed by the system during a transition between two non-equilibrium states.

On account of the positive semidefinite form of the dissipation function (5.1.14) (in connection with a stability criterion see section 6.1), the mechanical power equation can be converted into an inequality

$$\theta\pi - \dot{\varepsilon} \geq 0, \qquad \eta = \text{const} \tag{5.3.1}$$

Energy can be dissipated during the transition but the transition has to be reversible. After integrating in time, we obtain

$$\omega \geq \Delta\varepsilon \tag{5.3.2}$$

where ω is the work done on the system during the transition. Recall that π is the power per unit mass and temperature. If work is done on the system during the transition then, in the reverse transition, the system will perform work on the external agent. According to the usual convention (Landau and Lifshitz, 1958) $\omega < 0$ is the work done on the system and $|\omega| = -\omega$ will be the work done by the system so that inequality (5.3.2) can be alternatively written as

$$|\omega| \leq -\Delta\varepsilon \tag{5.3.3}$$

for the work done by the system. For a conservative process ($\phi = 0$), we have

$$|\omega|_{\max} = -\Delta\varepsilon, \qquad \eta = \text{const} \tag{5.3.4}$$

which, expressed in words, states that the maximum amount of work done by the system is equal to the decrease in the internal energy for those thermodynamic processes which are both reversible and non-dissipative. Although we have gone one step beyond the energy balance equation by introducing the notion of dissipation, the mechanical power equation in the energy representation cannot account for irreversible transitions.

Let us now consider the power equation in the entropy representation and ask what is the minimum power that can be supplied to the system as it makes a transition from one non-equilibrium state to another. In order to convert the power equation (5.2.1) into an inequality we have at our disposal the local form of the second law inequality. Moreover, since we assume this to be valid both in linear and nonlinear thermodynamics, the results to be obtained will have a general validity. If the thermodynamic process occurs at constant temperature and strain, the thermodynamic principle of the balance of power, as expressed by equation (5.2.1), reduces to

$$\theta\pi - \dot{\psi} = \theta\sigma_S, \qquad \dot{\mathbf{F}} = \mathbf{0},\ \dot{\theta} = 0 \tag{5.3.5}$$

and there are no other conservative external fields acting on the system. On account of the second law, we have the inequality

$$\theta\pi \geq \dot{\psi} \tag{5.3.6}$$

where the equality sign applies to a reversible process. During a reversible transition, we have therefore

$$\theta\pi_{\min} = \dot{\psi} \tag{5.3.7}$$

which is to be considered as the kinetic analogue of the principle of maximum work in section 1.5. The minimum power absorbed is clearly a negative quantity so it must correspond to the maximum power *output* that occurs only for reversible transitions. Therefore, the maximum power output appears as the rate at which the free energy decreases in time during a reversible process that is both isothermal and isochoric. For an irreversible process, part of the free energy will be absorbed in the irreversible nature of the transition. Consequently, both the work and the power output will not be at their maximum values.

We now ask for what is the maximum change in the power that can occur as the system undergoes a transition from one non-equilibrium state to another. In the energy representation the answer is clear enough: the maximum change in the power output is equal to the change in the rate at which the internal energy has decreased. This occurs when the transition is non-dissipative but, of course, reversible. When we want to consider the effect of irreversibility, we must consider the mechanical power equation in the entropy representation.

We are immediately confronted with the fact that we cannot use the criterion

$$\Delta\sigma_S \leq 0 \tag{5.3.8}$$

in nonlinear thermodynamics. In linear thermodynamics, the result is uninteresting since the power vanishes and the use of the criterion (5.3.8) leads to

$$\Delta\dot{\psi} \geq 0 \tag{5.3.9}$$

for a thermodynamic process at constant temperature and volume. However, as was shown in section 4.1, the Legendre transform of the entropy production ζ

will always satisfy the inequality

$$\Delta\zeta \leq 0 \tag{5.3.10}$$

The non-equilibrium thermodynamic potential ζ is defined by (4.1.44) and thus we can replace the entropy production in the power equation by

$$\sigma_S = -\chi_{Tj}\dot{\lambda}_j - \zeta \tag{5.3.11}$$

where the sum runs over all the velocities except those which are kept constant by the external constraints. Inserting (5.3.11) into the power equation (5.2.14), taking the Δ differential and using inequality (5.3.10), results in

$$\Delta\pi \geq \Delta(\sigma_S + \zeta - 2\phi) \tag{5.3.12}$$

When the equality sign applies, we obtain

$$\Delta\pi_{\min} = \Delta(\sigma_S - 2\phi) \tag{5.3.13}$$

The minimum change in the power is equal to the change in the difference between the entropy production and twice the dissipation. We obtain the unexpected result that the minimum change in the power is insensitive to whether the process is reversible or not. Now let us assume that the system is left to itself and there is no power input. Spontaneous, irreversible processes will arise that will bring the system back to equilibrium. The irreversible processes will occur such that the inequality

$$\Delta(\sigma_S + \zeta - 2\phi) \leq 0 \tag{5.3.14}$$

is satisfied. This means that all irreversible processes will cause the quantity $\sigma_S + \zeta - 2\phi$ to decrease. Inequality (5.3.14) is valid for arbitrarily large displacements from equilibrium. It answers the question of how the system will evolve when it has been brought to a non-equilibrium state by an external power source which has subsequently been turned off. The quantity $\sigma_S + \zeta - 2\phi$ will reach its minimum value not at equilibrium but in the linear range of thermodynamics. Other spontaneous processes will occur which will bring the system to equilibrium where each of the terms will vanish identically. We may thus conclude that inequality (5.3.14) is a nonlinear thermodynamic criterion of evolution, provided, of course, that the thermodynamic process admits a description in terms of scalar non-equilibrium thermodynamic potentials.

Another interesting result can be obtained when the entropy production is replaced by its Legendre transform in the power equation (5.2.1). Inserting (5.3.11) into the power equation results in

$$\pi + X_{Tj}\,\dot{\lambda}_j + \zeta = X_{Dj}\dot{\lambda}_j \tag{5.3.15}$$

where the sums extend over the N independent velocities. The power equation (5.3.15) will help us clear up a doubtful point concerning the expansion of the entropy production about a given non-equilibrium state (5.2.10). If we had treated the entropy production as a bilinear sum without regard to its functional dependency, then we would conclude that a linear term in the thermodynamic force is lacking. Specifically, let us consider that there is one velocity which is maintained constant by the external constraints $\dot{\lambda}_{N+1}$ and ask why the term $\chi_{T(N+1)}\dot{\lambda}_{N+1}$ did not appear in the expansion (5.2.10). It is apparent that if $\dot{\lambda}_{N+1}$ is to be kept constant then the force $\chi_{T(N+1)}$ must vary. However, in (5.2.10) we used explicitly the property that the entropy production is a first order homogeneous function of the velocities and expanded in terms of the independent velocities. Let us now see if we were correct in our assumption.

The term $X_{T(N+1)}\dot{\lambda}^0_{N+1}$ is added and subtracted from the left-hand side of the power equation (5.3.15). We then obtain the power equation

$$\pi + (\zeta - X_{T(N+1)}\dot{\lambda}^0_{N+1}) = -X_{T(N+1)}\dot{\lambda}^0_{N+1} - X_{Tj}\dot{\lambda}_j + X_{Dj}\dot{\lambda}_j$$
$$= \sigma_S - 2\phi \qquad (5.3.16)$$

where the sum is over the N independent velocities. Comparing equation (5.3.16) with the power equation (5.2.14) we conclude that the terms in the parenthesis are equal:

$$\zeta = X_{T(N+1)}\dot{\lambda}^0_{N+1} \qquad (5.3.17)$$

The total differential of (5.3.17) is

$$d\zeta = \dot{\lambda}^0_{N+1} dX_{T(N+1)} \qquad (5.3.18)$$

which, from (4.1.38) and (4.1.45), is seen to be equal to

$$d\zeta = -\dot{\lambda}_j dX_{Tj} \qquad (5.3.19)$$

This shows that if a term of the form $\chi_T\dot{\lambda}^0$ had appeared in (5.2.10) it would have meant that we were considering an expansion of both the entropy production and its Legendre transform about the given non-equilibrium stationary state. Obviously this is incorrect and (5.2.10) is the correct expansion for the entropy production.

Another interesting form of the power equation results when the Legendre transform of the dissipation function is introduced into the power equation in place of the dissipation function. Unlike the entropy production the total Legendre transform of the dissipation function does not vanish. This is due to the fact that ϕ is a second order homogeneous function of the independent velocities. In contrast, the non-equilibrium thermodynamic potentials σ_S and and π are first order homogeneous functions of all the velocities, which include the independent velocities as well as those which are maintained constant by the external constraints. Qualitatively speaking, dissipation is accounted for in the internal degrees of freedom of the system.

The total Legendre transform of the dissipation function ϕ defines the 'generating' function (Lavenda, 1974)

$$\Upsilon = (\partial_{\dot{\xi}_i}\phi)\dot{\xi}_i - \phi = -\chi_{Di}\dot{\xi}_i - \phi \qquad (5.3.20)$$

The functional dependencies of Υ are displayed by taking the total differential of (5.3.20)

$$d\Upsilon(\chi_{Di}, \chi_D) = [R^{-1}]_{ij}\chi_{D_j} d\chi_{Di} = -\dot{\xi}_i d\chi_{Di} \qquad (5.3.21)$$

which shows that the generating function is a second order homogeneous function of the dissipative forces. In generalised thermodynamics both the dissipation function ϕ and the generating function Υ have been confused with the entropy production, cf. equations (4.1.12) and (4.1.13).

Since

$$2\phi = -\chi_{Di}\dot{\xi}_i \qquad (5.3.22)$$

we obtain from the Legendre transform (5.3.20) that

$$\Upsilon(\chi_{D_1}, \chi_D) = \phi(\dot{\xi}, \xi) \tag{5.3.23}$$

This *numerical* equivalence permits us to write the mechanical power equation (5.2.14) in the more symmetric form

$$\sigma_S - \pi = \phi + \Upsilon \tag{5.3.24}$$

We stress the fact that although the two dissipation functions ϕ and Υ are numerically equal, they are not functionally equivalent. The distinction between functions which are numerically equal as opposed to functionally equivalent is crucial to our development of the variational principles of nonlinear thermodynamics in the next chapter. In fact, this chapter has paved the way for this development by (1) independently deriving the thermodynamic equations which will be shown to result from the condition of stationarity of the thermodynamic variational principle, and (2) elucidating the functional dependencies of the non-equilibrium thermodynamic potentials. It is not due to a mere oversight on the part of rational and generalised thermodynamics that variational principles were not formulated. Rather, it was due to the facts that the thermodynamic principles were not formulated and the non-equilibrium thermodynamic potentials were not functionally defined.

References

Casimir, H. B. G. (1945). On Onsager's principle of microscopic reversibility. *Rev. mod. Phys.*, **17,** 343–350.

Coleman, B. D. and Truesdell, C. (1960). On the reciprocal relations of Onsager. *J. chem. Phys.*, **33,** 28–31.

de Groot, S. and Mazur, P. (1962). *Non-equilibrium Thermodynamics.* North-Holland, Amsterdam.

Eringen, A. C. (1965). A unified theory of thermomechanical materials. *Int. J. engng Sci.*, **4,** 179–202.

Glansdorff, P. and Prigogine, I. (1971). *Thermodynamic Theory of Structure, Stability, and Fluctuations.* Wiley-Interscience, New York.

Landau, L. D. and Lifshitz, E. M. (1958). *Statistical Physics.* Pergamon Press, Oxford.

Lavenda, B. H. (1972). Concepts of stability and symmetry in irreversible thermodynamics I. *Found. Phys.*, **2,** 161–179.

Lavenda, B. H. (1974). Principles and representations of nonequilibrium thermodynamics. *Phys. Rev. A,* **9,** 929–943.

Machlup, S. and Onsager, L. (1953). Fluctuations and irreversible processes. II. Systems with kinetic energy. *Phys. Rev.*, **91,** 1512–1515.

Meixner, J. (1941). *Annln Phys.*, **39,** 333.

Meixner, J. (1943). *Z. phys. Chem. Abt. B,* **53,** 235.

Onsager, L. (1931). Reciprocal relations in irreversible processes I. *Phys. Rev.*, **37,** 405–426; Reciprocal relations in irreversible processes II. *Phys. Rev.*, **38,** 2265–2279.

Tisza, L. (1966). *Generalized Thermodynamics.* MIT Press, Cambridge, Mass.

6 Non-equilibrium Variational Principles

Thermodynamic variational principles offer both a unification and classification of the fundamental laws governing non-equilibrium processes. At the heart of all thermodynamic variational principles is the principle of least dissipation of energy (Onsager, 1931). As a result of Onsager's formulation of the principle of least dissipation of energy, the dissipation function has, by far, a more prominent and respectable role in thermodynamic variational principles than it has in the variational principles of mechanics (Routh, 1877).

The development of thermodynamic variational principles has been sporadic and in some cases incorrect. All the classical treatises of non-equilibrium thermodynamics bypass the subject completely. The author is aware of only one monograph that has attempted to incorporate variational principles in non-equilibrium thermodynamics (Gyarmati, 1970). Most of the work contained in this monograph can be considered as an outgrowth of Prigogine's (1954) interpretation of the principle of minimum entropy production as a Gaussian principle of least constraint (cf. section 4.1). Opinions regarding the relation between Prigogine's principle of minimum entropy production and Onsager's formulation of the principle of least dissipation of energy have been voiced by Ono (1961), Kirkaldy (1964) and Gyarmati (1965*a*, 1965*b*, 1965*c*, 1966). There is still no consensus of opinion on the subject and Gyarmati (1970) has even claimed that his formulation of Onsager's principle is more general than the original formulation, since variations are performed simultaneously with respect to the forces and fluxes!

The only other place where we hear of a type of thermodynamic variational principle is the Glansdorff–Prigogine construction of the 'local' potential (Prigogine, 1965). This, however, is not in the main track of thermodynamic variational principles. The local potential is a slight generalisation of the Glansdorff–Prigogine '$\delta_X P$' criterion in that it allows the phenomenological coefficient to be a function of the dependent variable. The local potential uses the Raleigh–Ritz procedure to determine the stationary state (Glansdorff and Prigogine, 1964). Since the local potential method has nothing to do with the thermodynamic variational principles to be described in this chapter, its assets are not discussed here but are left to the interested reader to consult the relevant literature.

Since so much of the thermodynamic variational formulation depends on the properties of the dissipation function, section 6.1 is devoted to such a discussion in terms of the dynamic analogues of the Le Châtelier and the Le Châtelier–Braun principles. Section 6.2 then develops the thermodynamic formulation of the principle of least dissipation of energy. There it is seen how the principle of minimum entropy production follows as a corollary from the principle of least dissipation of energy. With the realisation that the principle of least dissipation of energy is not a true variational principle in the classical mechanical sense, but rather a minimum principle, the analogy between it and the Gaussian principle of least constraint is drawn in section 6.3. The minimum principles of nonlinear thermodynamics are formulated in section 6.4 and in section 6.5 is provided a kinetic formulation of these principles that is, indeed, a true variational principle. Throughout this development the profound relationship between thermodynamic variational and statistical principles is stressed.

6.1 The dynamic Le Châtelier principle

Since the dissipation function ϕ is at the very heart of all thermodynamic variational principles, it is worth the time spent in discussing some of its properties.

The dissipation function ϕ is defined as

$$2\phi = -\boldsymbol{\chi}_D \cdot \boldsymbol{\iota} = R_{ij}\dot{\xi}_i\dot{\xi}_j \tag{6.1.1}$$

where the sum is over the N independent velocities that have been chosen such that they vanish in a given non-equilibrium stationary state. The existence of the dissipation function depends on whether the exactness conditions

$$(\partial_{\iota_j}\chi_{Di}) = (\partial_{\iota_i}\chi_{Dj}), \qquad R_{ij} = R_{ji} \tag{6.1.2}$$

are satisfied. These relations are commonly referred to as the Onsager symmetry relations which are necessarily valid in linear thermodynamics. But there is no guarantee that such relations will hold for nonlinear thermodynamic processes. So all what is to be said must always be prefaced by the remark that when a nonlinear thermodynamic process satisfies (6.1.2) then there exists a dissipation function, otherwise not.

Provided that the quadratic form (6.1.1) is non-singular,

$$D_N = |R_{ij}| \neq 0 \tag{6.1.3}$$

there will be an infinite number of linear (affine) transformations that will bring (6.1.1) into diagonal form. We will be interested in a unimodular transformation (that is of determinant unity), since we can then conclude that the transformation is volume preserving (cf. section 1.3). Applying a unimodular transformation to (6.1.1) results in the 'diagonal' representation of the dissipation function

$$2\phi = R_i\dot{\xi}_i'^2 \tag{6.1.4}$$

where the ξ''s are linear combinations of the original velocity variations. Since the transformation is unimodular, the discriminant of the quadratic form will be invariant, $D_N = R_1 R_2 \ldots R_N$. The principal minors can be written as

$$D_i = R_1 R_2 \ldots R_i = \frac{\partial(\chi_{D1}, \chi_{D2}, \ldots, \chi_{Di})}{\partial(\iota_1, \iota_2, \ldots, \iota_i)} \tag{6.1.5}$$

The eigenvalues of the quadratic form (6.1.4) are given by

$$R_i = -D_i/D_{i-1} = -\frac{\partial(\chi_{D1}, \ldots, \chi_{Di}, \iota_{i+1}, \ldots, \iota_N)}{\partial(\chi_{D1}, \ldots, \chi_{Di-1}, \iota_i, \ldots, \iota_N)}$$

$$= -(\partial_{\iota_i}\chi_{Di})_{\chi_{D1}, \ldots, \chi_{Di-1}, \iota_{i+1}, \ldots, \iota_N = 0} \tag{6.1.6}$$

since the partial derivative is evaluated at the stationary state where the independent velocities vanish together with the dissipative force.

A necessary but not a sufficient condition of stability is that the system dissipates positively. This means that when we consider the generalised Gibbs space, spanned by all the independent velocities, the dissipation function must be represented by an ellipsoid centred on the origin. The conditions are

(1) Ellipsoid: all $R_i > 0$.
(2) Paraboloid: all $R_i \geq 0$, with one or more $R_i = 0$.
(3) Hyperboloid: one or more $R_i < 0$.

In the next chapter it will be seen that (1) is a necessary but not sufficient condition of stationary state stability. On the other hand, (2) is a sufficient condition for instability. In other words, a single internal source of power (that is, negative dissipation) is sufficient to ultimately make the system unstable. Condition (2) is a criterion of meta-stability or what generalised thermodynamics refers to as a 'marginally' stable state (cf. section 4.2).

The foregoing observations can be unified into a single statement that constitutes a dynamic Le Châtelier principle. For simplicity, we treat the case of two independent velocities and then generalise to the N dimensional case. The necessary condition of stability (1) requires

$$R_1 = -(\partial_{\iota_1}\chi_{D1})_{\iota_2 = 0} > 0 \quad \text{and} \quad R_2 = -(\partial_{\iota_2}\chi_{D2})_{\chi_{D1} = 0} > 0 \tag{6.1.7}$$

The dynamic Le Châtelier principle states that the dissipative force is a decreasing function of its conjugate velocity in all locally stable systems. Consider the second condition in (6.1.7); it can be written as

$$R_2 = \frac{(\partial_{\iota_2}\chi_{D1})^2_{\iota_1 = 0}}{(\partial_{\iota_1}\chi_{D1})_{\iota_2 = 0}} - (\partial_{\iota_2}\chi_{D2})_{\iota_1 = 0} > 0 \tag{6.1.8}$$

From (6.1.7) and (6.1.8) we conclude that

$$(\partial_{\iota_2}\chi_{D2})_{\iota_1 = 0} \leq (\partial_{\iota_2}\chi_{D2})_{\chi_{D1} = 0} < 0 \tag{6.1.9}$$

We can immediately generalise to the N-dimensional case; using absolute values we have

$$|(\partial_{\iota_i}\chi_{Di})_{\iota_j = 0}| \geq |(\partial_{\iota_i}\chi_{Di})_{\chi_{D1} = 0}| \geq \ldots \geq |(\partial_{\iota_i}\chi_{Di})_{\chi_{D1}, \ldots, \chi_{Di-1} = 0}| \tag{6.1.10}$$

This is the generalised form of the dynamic Le Châtelier principle: it states that when a locally stable system is displaced from a given stationary state by a variation in the velocity ι_i, the system responds by changing its conjugate force χ_{Di}. The response is largest when all the $N-1$ other independent velocities are maintained at their zero stationary state value. Moreover, it decreases as the other independent velocities are liberated. Qualitatively speaking, we say that as the effect of the initial perturbation becomes more evenly distributed among the internal degrees of freedom, the response decreases.

Having obtained a dynamic counterpart to the equilibrium Le Châtelier principle, we now inquire as to whether there exists a dynamic counterpart of the Le Châtelier – Braun principle. We ask for what is the effect of the spontaneous behaviour of a secondary velocity ι_2 on the primary force χ_{D1} when the system is perturbed from a given stationary state by creating a velocity ι_1 and then releasing all constraints on both velocities ι_1 and ι_2. In the presence of a constraint on ι_2, which maintains it at its stationary state zero value, perturbing the system will cause a change in the velocity ι_1. Consequently, the system dissipation will increase by an amount

$$d\phi = -\chi_{D1} d\iota_1 \tag{6.1.11}$$

Then, according to the dynamic principle of Le Châtelier, all spontaneous processes will lead to a decrease in the system dissipation, making it tend to a minimum. Here, we have the most primitive statement of the principle of least dissipation of energy. The change in ι_1, namely $\dot{\xi}_1$, tends to decrease the conjugate dissipative force

$$(\partial_{\iota_1}\chi_{D1})_{\iota_2=0} < 0 \tag{6.1.12}$$

In order for the dissipation function to decrease, both χ_{D1} and $d\iota_1$ must be of the same sign, say positive. Now the creating of a velocity ι_1 will cause a change in the secondary force χ_{D2} which is related to χ_{D1} by

$$\chi_{D2} = (\partial_{\chi_{D1}}\chi_{D2})_{\iota_2=0}\,\chi_{D1} \tag{6.1.13}$$

When the constraints are released, both χ_{D2} and $d\iota_2$ must be of the same sign in order for the dissipation to decrease further. This is expressed by

$$\operatorname{sgn}(d\iota_2) = \operatorname{sgn}\left[(\partial_{\chi_{D1}}\chi_{D2})_{\iota_2=0}\right] \tag{6.1.14}$$

since we have taken χ_{D1} to be positive. Calculating the effect of the secondary velocity ι_2 on the primary force, we obtain

$$(d\chi_{D1})_{\iota_1=0} = (\partial_{\iota_2}\chi_{D1})_{\iota_1=0}\,d\iota_2 \tag{6.1.15}$$

From (6.1.14) and (6.1.15) it is apparent that

$$\operatorname{sgn}\left[(d\chi_{D1})_{\iota_1=0}\right] = \operatorname{sgn}\left[(\partial_{\iota_2}\chi_{D1})_{\iota_1=0}\cdot(\partial_{\chi_{D1}}\chi_{D2})_{\iota_2=0}\right] \tag{6.1.16}$$

Evaluating (6.1.16) with the first stability criterion in (6.1.7), we conclude that

$$(d\chi_{D1})_{\iota_1=0} < 0 \tag{6.1.17}$$

Inequality (6.1.17) constitutes the essence of the Le Châtelier – Braun principle: the effect of the spontaneous change in the secondary velocity ι_2 is to diminish the primary dissipative force χ_{D1} caused by the initial perturbation of the system from the given stationary state.

These criteria of stationary state stability, together with the fact that the dissipation function is a non-classical thermodynamic potential, i.e. it is a second order homogeneous function, lend themselves very nicely to the construction of a non-equilibrium thermodynamic variational principle known as the principle of least dissipation of energy.

6.2 The principle of least dissipation of energy

Rayleigh (1873) constructed the dissipation function to account for damping effects in mechanical systems. In terms of the variational principles of classical mechanics, the dissipation function has an awkward role. Although it has the same form of the '*vis viva*', the dissipation function does not enter into the variational principle itself but rather its derivative, with respect to the velocity, is added as a non-conservative force to the Euler – Lagrange equations that result from the stationarity condition of the conservative variational principle (Routh, 1877).

The dissipation function has a far more respectable role in non-equilibrium thermodynamic variational principles, as Onsager has pointed out on a number of occasions (Onsager, 1931; Onsager and Machlup, 1953: Machlup and Onsager, 1953). In his original formulation, Onsager (1931) applied the principle of least dissipation of energy to the conduction of heat in a closed system. This is undoubtedly the most questionable application of the principle of least dissipation of energy that can be made, since it requires the dissipation function to have the form

$$2\rho\phi = \boldsymbol{q}\mathbf{K}^{-1}\boldsymbol{q} \tag{6.2.1}$$

where $\mathbf{K}$ is the thermal conductivity tensor. As already mentioned in section 2.3, the heat flux vector is not uniquely defined and only its divergence has a real physical significance. Moreover, since $\boldsymbol{q}$ is not a rate variable, we cannot establish the symmetry of $\mathbf{K}$ from considerations involving detailed balancing at equilibrium. However, since the divergence of $\boldsymbol{q}$ is what is important and we assume the elements of $\mathbf{K}$ to be constants, then the antisymmetric part of $\mathbf{K}$ will have no observable physical consequence and we can put it equal to zero (Casimir, 1945) (cf. section 2.3).

If we accept this argument then we can consider the dissipation function to be the potential of the relevant dissipative force, the gradient of temperature. The heat conduction power equation (5.1.7) reduces to the very simple form

$$2\phi(\boldsymbol{q}, \boldsymbol{q}) = -\frac{1}{\rho\theta^2}\,\boldsymbol{q}\cdot\boldsymbol{g} \tag{6.2.2}$$

We also know that the energy balance in this case is

$$\dot{\varepsilon} = -\frac{1}{\rho}\nabla\cdot\boldsymbol{q} \tag{6.2.3}$$

while the entropy balance equation is given by

$$\dot{\eta} + \frac{1}{\rho}\nabla\cdot(\boldsymbol{q}/\theta) = \sigma_S \tag{6.2.4}$$

Using the relation $\dot{\varepsilon} = \theta \dot{\eta}$, we find that the specific entropy production will be equal to (6.2.2) so that the following (linear) thermodynamic principle applies:

$$2\phi(\boldsymbol{q}, \boldsymbol{q}) = \sigma_S(\boldsymbol{q}) \tag{6.2.5}$$

This is to say that, in linear thermodynamics, the thermodynamic principle of the balance of power reduces to the statement that the entropy produced appears as the dissipated energy. It was first formulated by Onsager (1931).

Onsager then went on to state that the linear thermodynamic variational principle is

$$\phi(\boldsymbol{q}, \boldsymbol{q}) - \sigma_S(\boldsymbol{q}) = \min \tag{6.2.6}$$

Why this form? Besides giving the correct results, it has been shown to follow from the constrained principle of least dissipation of energy (Lavenda, 1974)

$$\phi(\boldsymbol{q}, \boldsymbol{q}) = \min \tag{6.2.7}$$

which is subject to the constraint imposed by the thermodynamic principle (6.2.5). Why does (6.2.5) impose a constraint on (6.2.7)?

The reason is that we are dealing with a so-called 'conditioned' minimum and not a 'free' minimum of the dissipation function. This is to say, we are not merely interested in finding the value of the heat flux for which the dissipation becomes a minimum, but we are asked at the same time to consider only that value of the heat flux for which the (linear) thermodynamic principle of the balance of power (6.2.5) is simultaneously fulfilled.

Lagrange suggested a simple and straightforward way of handling such a problem. We can introduce the 'constraint' (6.2.5) explicitly into the variational principle (6.2.7) by multiplying $(2\phi - \sigma_S)$ by an undetermined multiplier λ and adding it to (6.2.7). We then obtain

$$\phi + \lambda(2\phi - \sigma_S) = \min \tag{6.2.8}$$

which is the same as (6.2.7), since all we have done is to add zero. Expression (6.2.8) is now a *free* variational principle. Using the 'convention' that the heat flux vector is to be varied for a prescribed temperature distribution (Onsager, 1931), we find the stationarity condition of (6.2.8) to be

$$[\partial_{\boldsymbol{q}}\phi + \lambda(2\partial_{\boldsymbol{q}}\phi - \partial_{\boldsymbol{q}}\sigma_S)].\delta\boldsymbol{q} = 0 \tag{6.2.9}$$

Since variations in the heat flux vector are completely arbitrary, the only way to satisfy (6.2.9) is to require that its coefficient vanish. This then allows us to determine the Lagrange undetermined multiplier λ. Taking the scalar product of

$$\partial_{\boldsymbol{q}}\phi - \lambda(2\partial_{\boldsymbol{q}}\phi - \partial_{\boldsymbol{q}}\sigma_S) = \boldsymbol{0} \tag{6.2.10}$$

with the heat flux vector, and using the properties that the dissipation function and entropy production are second and first order homogeneous functions of $\boldsymbol{q}$, we obtain

$$2(1 + 2\lambda)\phi - \lambda\sigma_S = 0 \tag{6.2.11}$$

In order for (6.2.11) to agree with the thermodynamic principle (6.2.5), $\lambda = -1$. Introducing this value of λ into (6.2.8) yields $\sigma_S - \phi = \max$, or equivalently

(6.2.6), since the extremum property of the variational principle depends solely on the dissipation function.

Noting (6.2.2) and (6.2.5), we can write the variational principle (6.2.6) as

$$\rho\phi(\boldsymbol{q}, \boldsymbol{q})+\frac{1}{\theta^2}\boldsymbol{q}\cdot\boldsymbol{g} = \min \tag{6.2.12}$$

On account of the fact that variations in the heat flux are now free, the condition of stationarity is

$$\rho\boldsymbol{\chi}_D-\frac{1}{\theta^2}\boldsymbol{g} = 0 \tag{6.2.13}$$

where

$$\partial_{\boldsymbol{q}}\phi = -\boldsymbol{\chi}_D \tag{6.2.14}$$

The scalar product of the stationarity condition (6.2.13) with the heat flux vector gives back equation (6.2.2). From (6.2.1), (6.2.13) and (6.2.14) we obtain the Fourier heat conduction law in the form

$$\boldsymbol{q} = \mathbf{K}\cdot\boldsymbol{\nabla}(1/\theta) \tag{6.2.15}$$

Finally, the minimum property of the variational principle follows from the facts that the dissipation function is a second order homogeneous function and the stability condition that the eigenvalues of the thermal conductivity tensor are all positive.

Although we have treated a relatively simple, albeit somewhat questionable, application of the principle of least dissipation of energy, it serves to illustrate the characteristic features of the thermodynamic variational principle. We shall now comment on some of these features.

First, the thermodynamic variational principle (6.2.6) is not a *true* variational principle in the classical mechanical sense of the meaning. Simply speaking, it does not employ the calculus of variations. Rather, it has the form of a minimum principle. In analogy with the variational principles of classical mechanics, it would be closer to d'Alembert's principle than to Hamilton's principle. Although the two principles are mathematically equivalent, d'Alembert's principle makes an independent statement at each instant of the motion, whereas Hamilton's principle considers the motion as a whole, i.e. during a given time interval.

Secondly, we know that in classical mechanics, d'Alembert's principle can be transformed into a genuine minimum principle, known as Gauss's principle of least constraint, by choosing a special form of variation. We recall from section 4.1 that this was the motivation for Prigogine (1954) to derive the principle of minimum entropy production using the nostrum of varying simultaneously and independently the fluxes and forces. Our illustration of the principle of least dissipation of energy shows that the entropy production is a first order homogeneous function of the velocities and, as such, cannot be a candidate for a variational principle.

We can, however, obtain the principle of minimum entropy production as a corollary to the variational principle of least dissipation of energy. This can be shown by introducing equation (6.2.5) into the constrained variational principle (6.2.7). We then obtain

$$\sigma_S(\boldsymbol{q}) = \text{min value} \tag{6.2.16}$$

We strongly emphasise the fact that there is a definite distinction between the minimum value of a function and the condition that the function be a minimum in the variational sense.

Thirdly, if we had introduced the thermodynamic principle (6.2.5) into (6.2.6) rather than into (6.2.7), we would have obtained the absurd result: $\sigma_S(\boldsymbol{q}) = \max$ value. This is obviously incorrect, since we have already used the thermodynamic principle (6.2.5) to convert the constrained variational principle (6.2.7) into the free variational principle (6.2.6). Apparently, Onsager (1931) did not recognise the fact that (6.2.6) is the free variational principle corresponding to the constrained variational principle (6.2.7), and erred in the application of the principle of least dissipation of energy to the case of stationary heat flow. Onsager's argument ran as follows: consider the case of stationary heat flow

$$\nabla \cdot \boldsymbol{q} = 0 \tag{6.2.17}$$

and consequently $\dot{\eta} = 0$, cf. equations (6.2.3) and (6.2.4). Then in the case of stationary heat flow, the variational principle (6.2.6) reduces to

$$\phi(\boldsymbol{q}, \boldsymbol{q}) - \dot{\eta}^*(\boldsymbol{q} \cdot \boldsymbol{n}) = \min \tag{6.2.18}$$

where, to use Onsager's notation, we have denoted $\dot{\eta}^*(\boldsymbol{q} \cdot \boldsymbol{n})$ as the entropy current and $\boldsymbol{n}$ is the unit normal directed outward from the surface of the system.

For stationary heat flow, the thermodynamic principle (6.2.5) reduces to

$$\dot{\eta}^*(\boldsymbol{q} \cdot \boldsymbol{n}) = 2\phi(\boldsymbol{q}, \boldsymbol{q}) \tag{6.2.19}$$

which obviously can be written as

$$2[\phi(\boldsymbol{q}, \boldsymbol{q}) - \dot{\eta}^*(\boldsymbol{q} \cdot \boldsymbol{n})] = -\dot{\eta}^*(\boldsymbol{q} \cdot \boldsymbol{n}) \tag{6.2.20}$$

If (6.2.18) is now multiplied by two and (6.2.20) is inserted, we would come to the erroneous conclusion

$$\dot{\eta}^*(\boldsymbol{q} \cdot \boldsymbol{n}) = \max \text{ value} \tag{6.2.21}$$

This is to say that the entropy production tends to its maximum value under the condition of stationary heat flow. Furthermore, if certain restrictions are imposed on the flow, 'like cracks in a crystal', we would get a new heat flux vector $\boldsymbol{q}'$ for which

$$\dot{\eta}^*(\boldsymbol{q} \cdot \boldsymbol{n}) \geq \dot{\eta}^*(\boldsymbol{q}' \cdot \boldsymbol{n}) \tag{6.2.22}$$

Thus, Onsager (1931) concluded that defects that alter the heat flow can only decrease the entropy production under stationary heat flow conditions, or cause no change. This is indeed a surprising result, since we would anticipate that a decrease in symmetry would increase the entropy production instead of lowering it.

Alternatively, if the thermodynamic principle of the balance of power (6.2.19) is introduced into the principle of least dissipation of energy (6.2.7), we obtain

$$\dot{\eta}^*(\boldsymbol{q} \cdot \boldsymbol{n}) = \min \text{ value} \tag{6.2.23}$$

and not (6.2.21). This means that it is incorrect to introduce the thermodynamic principle (6.2.19) *twice* into the constrained variational principle (6.2.7), first as a constraint and second to obtain a criterion for the entropy current. This being the case, for any other heat flux $\boldsymbol{q}'$, due to say crystal imperfections, we would get

$$\dot{\eta}^*(\boldsymbol{q}\cdot\boldsymbol{n}) \leq \dot{\eta}^*(\boldsymbol{q}'\cdot\boldsymbol{n}) \tag{6.2.24}$$

Alterations of the heat flux vector, due to say imperfections in the conducting material, can only *increase* the entropy production, or cause no change, under stationary heat flow conditions.

6.3 Gauss's principle in non-equilibrium thermodynamics

The correct thermodynamic analogue of Gauss's principle of least constraint was given by Onsager and Machlup (1953)—but for a different intention. These authors were concerned with the determination of the probability for the regression of a fluctuation from a given non-equilibrium state and interpreted the variational equations (5.2.21) as stochastic differential equations with χ_E as a Gaussian, random force.

Gauss's principle can be used to determine the error between the hypothetical value of a function and the observed value, in perfect analogy with the 'method of least squares'. And if it were not for the random force there would be no error, so that the non-equilibrium process would always follow the path described by

$$R_{ij}\dot{\xi}_j + S_{ij}\xi_j = 0 \tag{6.3.1}$$

or

$$\bar{\bar{\chi}}_D = \chi_T; \quad \bar{\bar{\chi}}_E = 0 \tag{6.3.2}$$

where the double bar symbol denotes the average value of the function. Of course, what we are saying applies only to isolated irreversible processes that occur in linear thermodynamics. But by the association of the error with the cause of random fluctuations, we can begin to appreciate the profound analogy between thermodynamic variational and statistical principles (Lavenda, 1977).

Thus, we can define the 'error' that an irreversible process will choose a path other than that described by equations (6.3.1) as the sum

$$\mathscr{G} = \frac{1}{2R_i}(\chi_{Di} - \bar{\bar{\chi}}_{Di})^2 = \frac{1}{2R_i}(R_i\dot{\xi}_i - \chi_{Ti})^2 = \min \tag{6.3.3}$$

in the diagonal representation. We have required that the error be a minimum. The eigenvalues of the generalised resistance matrix can be interpreted as weighting factors that provide an estimate of reliability of the observations. Furthermore, the thermodynamic analogue of Gauss's principle (6.3.3) brings out the significance of the convention which varies the velocities for a prescribed configuration of the system in the thermodynamic variational principle. In terms of Gauss's principle (4.1.15), the thermodynamic force would correspond to the 'impressed' force; that is, the external force acting on a material object cannot be varied. This is equivalent to the condition of maintaining a prescribed configuration, since the thermodynamic force is a function of the generalised displacements and not on their rates of change. Moreover, in Gauss's principle, variations are performed with respect to the accelerations, whereas in the thermodynamic analogue, variations are taken with respect to the independent velocities. In classical mechanics, the impressed force will cause deviations in the velocity, whereas in non-equilibrium thermodynamics, the thermodynamic force will cause deviations in the generalised displacements. Then interpreting $\mathscr{G}$ as the constraint on the path of the irreversible process, we obtain the condition of stationarity as

$$R_i\dot{\xi}_i + S_i\xi_i = 0 \tag{6.3.4}$$

for which $\mathscr{G}$ assumes its absolute minimum and the irreversible process is free of constraints.

The thermodynamic constraint $\mathscr{G}$ takes on a particularly interesting form when we express it in terms of the non-equilibrium thermodynamic potentials (Onsager and Machlup, 1953)

$$\mathscr{G} = (\phi + \Upsilon - \sigma_S) = \min \tag{6.3.5}$$

where we have made use of (6.3.2) in identifying the entropy production. The function $\mathscr{G}$ has been called the Onsager–Machlup function (Tisza and Manning, 1957) and it has been used widely in the literature in connection with the calculation of the probability for a path of a non-equilibrium process (Onsager and Machlup, 1953; Machlup and Onsager, 1953; Tisza and Manning, 1957; Lavenda, 1977). Its relation to a Gaussian principle of mechanics has been emphasised by Gyarmati (1965*a*, 1965*b*, 1965*c*, 1970).

Inspired by the question of Prigogine (1954) as to whether there exists a Gaussian principle of least constraint in non-equilibrium thermodynamics, and misled by the Prigogine nostrum of varying simultaneously the fluxes and forces, Gyarmati (1970) proposed expression (6.3.5) as a generalisation of the thermodynamic analogue of Gauss's principle, for which the forces and fluxes are to be varied simultaneously and independently! Furthermore, Gyarmati (1970) claimed that the proposed 'force representation' of the Gaussian principle (6.3.5) is 'self-explanatory, namely σ_S is a symmetric bilinear form in terms of forces and fluxes'. Yet, if $\mathscr{G}$ is to be varied with respect to the forces at fixed constant values of the velocities, there results

$$\delta\mathscr{G} = [\boldsymbol{\imath}\cdot\delta\boldsymbol{\chi}_T + \delta\Upsilon] = 0 \tag{6.3.6}$$

since the dissipation function has a mute role in the variation with respect to the forces. It will now be appreciated that the scalar product vanishes on account of the non-equilibrium Gibbs–Duhem relation (4.1.35) which renders (6.3.6) meaningless. The vanishing of the scalar product in (6.3.6) means that all the thermodynamic forces cannot be varied simultaneously. In other words, the non-equilibrium Gibbs–Duhem relation (4.1.35) preserves the first order homogeneous property of the entropy production. It can also be given a geometrical significance, in that the velocity vector is perpendicular to any possible virtual variation of the thermodynamic force or equivalently to that of the generalised displacement vector.

Instead of varying $\mathscr{G}$ with respect to the thermodynamic forces, Gyarmati should have considered the force variation of

$$\mathscr{G}^* = (\phi + \Upsilon - \zeta) = \min \tag{6.3.7}$$

where we recall that ζ is the Legendre transform of the entropy production defined in (4.1.44). The distinction between $\mathscr{G}$ and $\mathscr{G}^*$, or for that matter (6.2.6) and

$$\Upsilon - \zeta = \min \tag{6.3.8}$$

is that $\mathscr{G}$ is varied with respect to the velocities at constant values of the thermodynamic forces, whereas $\mathscr{G}^*$ is varied with respect to the independent thermodynamic forces at constant values of the velocities. Although we do not dispute the fact that the results are identical, since the generalised resistance matrix is invertible, the physical significance of varying the independent thermodynamic forces is far from being clear. Moreover, it goes against the spirit of the Gaussian principle of least constraint where the impressed force is held fixed. It would also appear that (6.2.6) and (6.3.8) are analogous to the extremum property, say, of the internal energy and one of its Legendre transforms in equilibrium thermodynamics. Then whether one should use (6.2.6) or (6.3.8) would depend upon the type of experiment that is contemplated.

All that we have said so far concerning non-equilibrium extremum principles applies strictly to linear thermodynamics. In nonlinear thermodynamics the difference is that the first variations of the entropy production and power functions do not vanish. This will be recalled as the major obstacle confronted in generalised thermodynamics. In the next section the variational principles of nonlinear thermodynamics are formulated.

6.4 Variational principles of nonlinear thermodynamics

The extension of the principle of least dissipation of energy (6.2.7) to nonlinear thermodynamics is formally analogous to the expansion of the '*vis viva*' about a steady state of the motion (Routh, 1877). The results will not differ appreciably from those obtained in the previous two sections. We are not, however, requiring the non-equilibrium thermodynamic potentials to have extrema anywhere but at equilibrium. Rather, we are analysing the stability of small motions about a given non-equilibrium stationary state for which the principle of least dissipation of energy (6.2.7) is still applicable. Moreover, we shall find that (6.2.7) is insufficient to guarantee the stability of a given non-equilibrium stationary state. This means that the stability properties of a non-equilibrium stationary state cannot be analysed in terms of any single non-equilibrium thermodynamic potential, and herein lies the 'rub' in generalised thermodynamics.

Our interpretation of the principle of least dissipation of energy will henceforth be mechanical in nature. We are therefore excluding the questionable case of (6.2.1). The dissipation function will thus be a second order homogeneous function of the rate variables (5.2.15). But we can no longer conclude, in nonlinear thermodynamics, that the exactness conditions (6.1.2) will necessarily be satisfied by every thermodynamic process. The only admissible thermodynamic process will be those that do not destroy the exactness conditions (6.1.2) and

$$(\partial_{\alpha_i}\chi_{Tj}) = (\partial_{\alpha_j}\chi_{Ti}) \qquad S_{ji} = S_{ij} \tag{6.4.1}$$

in the case where the variational equations are described in terms of half-degrees of freedom. In the case of single degree of freedom variational equations, we would require $M_{ij} = M_{ji}$ in addition to the other two sets of exactness conditions (cf. section 7.2).

The mechanical power equation (5.2.14) is introduced into the constrained variational principle of least dissipation of energy

$$\phi(\dot{\xi}, \dot{\xi}) = \min \tag{6.4.2}$$

by the usual procedure of the Lagrange undetermined multiplier. We then obtain the free variational principle

$$\mathscr{L} = (\pi - \sigma_S + \phi) = \min \tag{6.4.3}$$

The thermodynamic Lagrangian $\mathscr{L}$ is centred on a given non-equilibrium stationary state, where it is expanded in series as (Lavenda, 1972)

$$\mathscr{L} = \mathscr{L}^0 + \mathscr{L}_1 + \mathscr{L}_2 + \ldots \tag{6.4.4}$$

It will be appreciated that expansion (6.4.4) is completely analogous to the classical mechanical expansion of the Lagrangian about a given steady state of the motion (Routh, 1877). If the reference state of the mechanical system is at rest then $\mathscr{L}^0 = \mathscr{L}_1 = 0$ and $\mathscr{L} = \mathscr{L}_2$ would describe small vibrations about a point of 'equilibrium'. Moreover, since ϕ is a second order quantity it will enter only in the quadratic term of the thermodynamic Lagrangian $\mathscr{L}_2$. This is only another way of saying that the zero and first order Lagrangians determine the conditions of the stationary state (which in classical mechanics would be a state of steady motion), while the stability of this state is still governed by $\mathscr{L}_2$ (Routh, 1877, especially ch. 4, art. 4). We can thus obtain the equations of small motions about a non-equilibrium stationary state in terms of a thermodynamic variational principle.

The expansions for the entropy production (5.2.10) and the power (5.2.12) about a given non-equilibrium stationary state are introduced into the thermodynamic Lagrangian (6.4.3). Setting $\mathscr{L}^0 = 0$, we obtain the condition for the formation of the non-equilibrium stationary state (5.2.17). The condition that the first order thermodynamic Lagrangian must vanish, $\mathscr{L}_1 = 0$, gives the stationary force balance law (5.2.18). We are then left with the second order term

$$\mathscr{L}_2 = (\delta^2\pi - \delta^2\sigma_S + \phi) = \min \tag{6.4.5}$$

which determines the stability of the stationary state. Using the convention that the independent velocities are to be varied at a constant value of the thermodynamic force, we obtain the condition of stationarity for $\mathscr{L}_2$ as

$$\delta\mathscr{L}_2 = 0\colon \quad \{\partial_\iota[\delta^2\pi - \delta^2\sigma_S + \phi]\}\cdot\delta\dot{\xi} = 0 \tag{6.4.6}$$

or equivalently, in terms of a variational force balance relation,

$$\chi_E = \chi_T - \chi_D \tag{6.4.7}$$

In the case where the thermodynamic force is a function of the generalised displacements only, the stationarity condition for $\mathscr{L}_2$ gives the variational equations of half-degrees of freedom (5.2.21). Alternatively, when the thermodynamic force is a function of both the generalised displacements and accelerations, (6.4.6) or (6.4.7) give the single degree of freedom variational equations (5.2.31).

The scalar product of the variational force balance equation (6.4.7) with the velocity vector gives the excess power equation $\delta^2\pi = \delta^2\sigma_S - 2\phi$ (5.2.19). Introducing this local, thermodynamic principle of the balance of power into the principle of least dissipation of energy (6.4.2) yields

$$\delta^2\sigma_S - \delta^2\pi = \text{min value} \tag{6.4.8}$$

which, expressed in words, states that the difference between the excess entropy production and the power tends to a minimum in a locally stable stationary state. Furthermore, the minimum property of the thermodynamic Lagrangian $\mathscr{L}_2$, (6.4.5), is determined solely by the positive definite, or a worst semidefinite, form of the dissipation function

$$\delta^2\mathscr{L}_2 \geq 0: \quad \partial^2_{\iota_i}\phi = R_i \geq 0 \tag{6.4.9}$$

We recall from section 6.1 that conditions (6.4.9) are necessary but not sufficient criteria of non-equilibrium stationary state stability.

6.5 Kinetic formulation of thermodynamic variational principles

So far, this chapter has discussed thermodynamic variational principles that, technically speaking, are actually extremum principles since they do not require the calculus of variations. A true variational principle can, however, be constructed from the thermodynamic analogue of Gauss's principle (6.3.5) in the event that the entropy current vanishes. Expression (6.3.5) can then be written as

$$\mathscr{G} = (\phi + \Upsilon - \dot{\eta}) \tag{6.5.1}$$

which we proceed to integrate in time between the limits $t = t_1$ and $t = t_2$

$$\mathscr{A} = \int_{t_1}^{t_2} (\phi + \Upsilon)\mathrm{d}t - [\partial_\alpha \eta \cdot \xi]_{t_1}^{t_2} \tag{6.5.2}$$

The second term in (6.5.2) has no role in the variational principle, since we vary between definite limits. The thermodynamic action $\mathscr{A}$ is effectively, then, the time integral of the sum of dissipation functions (Lavenda, 1977).

The stationary value of the thermodynamic action (6.5.2) is determined in the same way as the classical mechanical action, namely

$$\delta\mathscr{A} = 0: \quad (\mathrm{d}_t\partial_{\dot{\xi}}\phi - \partial_\xi\Upsilon)\cdot\delta\xi = 0 \tag{6.5.3}$$

Since we are considering the case where the external force variation is zero, we have used the fact that $\chi_D = \chi_T$; in contrast, see (6.5.11) below. Here, the stationarity condition (6.5.3) does not result from an expansion of the thermodynamic action about a given stationary state, since the state in question is equilibrium. Since we have obtained no stationary conditions, we are dealing with internal, irreversible transformations that are going on in an isolated system. This coincides with linear thermodynamics and is to be contrasted with (6.5.9) below, which is the thermodynamic action in nonlinear thermodynamics.

The stationarity condition (6.5.3) gives the equations of motion

$$R^2_{ij}\ddot{\xi}_j - S^2_{ij}\xi_j = 0, \qquad i = 1, 2, \ldots, N \tag{6.5.4}$$

which unhappily do not coincide with either set of variational equations (5.2.21) or (5.2.31) for vanishing values of the external force variations. However, it will be appreciated that the equations of motion (6.5.4) can be written as the product of two factors

$$(R_{ij}\mathrm{d}_t + S_{ij})(R_{ij}\mathrm{d}_t - S_{ij})\xi_j = 0, \quad i = 1, 2, \ldots, N \tag{6.5.5}$$

The solutions of these equations are superpositions of the solutions of the equations obtained from the factors

$$R_{ij}\dot{\xi}_j + S_{ij}\xi_j = 0, \qquad i = 1, 2, \ldots, N \tag{6.5.6}$$

$$R_{ij}\dot{\xi}_j - S_{ij}\xi_j = 0, \qquad i = 1, 2, \ldots, N \tag{6.5.7}$$

Although we recognise the first set of equations as the variational equations of half-degrees of freedom (5.2.21) in the case of vanishing external force variations, how can the variation principle distinguish between this set of equations and their mirror images (6.5.7)?

Only one set of variational equations, obtained from the stationarity condition of the thermodynamic action, will lead to a minimum of the thermodynamic action, or equivalently of $\mathscr{G}$. Since $\mathscr{G}$ is a sum of essentially positive terms it must have a minimum somewhere. The solutions of equations (6.5.6) are exponentially decreasing functions of time. In the diagonal representation, the solutions are of the form $\exp(-\beta_i t)$, where $\beta_i = S_i/R_i$. Using this set of solutions to evaluate the thermodynamic action (6.5.2), we obtain

$$\mathscr{A}_{\min} = \left[\int_{t_1}^{t_2} \mathscr{G}\,\mathrm{d}t\right]_{\min} = 0 \tag{6.5.8}$$

Consequently, the fact that the thermodynamic action must be a minimum selects the correct set of variational equations that are obtained, along with their mirror images, from the condition that the thermodynamic action be stationary.

Before considering some of the general features of the thermodynamic variational principle (6.5.2), let us generalise it to include nonlinear thermodynamic processes. The variational principle of nonlinear thermodynamics is

$$\mathscr{A} = \int_{t_1}^{t_2} (\phi + \Upsilon + \pi)\mathrm{d}t - [\partial_{\boldsymbol{a}}\eta\cdot\boldsymbol{a}]_{t_1}^{t_2} \tag{6.5.9}$$

Let us note carefully that the power cannot be brought out from under the integral sign as we have done for $\dot{\eta}$, since π is, in general, not the total differential of the generalised work function. In other words, the external force is, in general, an explicit function of time.

The zero and first order terms of the thermodynamic action, or equivalently of the thermodynamic Lagrangian, yield the stationary state condition and the stationary state force balance law, respectively. There then remains

$$\mathscr{A}_2 = \int_{t_1}^{t_2} (\phi + \Upsilon + \delta^2\pi)\mathrm{d}t - [\tfrac{1}{2}\delta^2\eta]_{t_1}^{t_2} \tag{6.5.10}$$

which is required to be a minimum. The generating function is now given by

$$2\Upsilon = \frac{1}{R_i}\chi_{Di}^2 = \frac{1}{R_i}(\chi_{Ti} - \chi_{Ei})^2 \tag{6.5.11}$$

in contrast to linear thermodynamics where the external force variation vanishes and we have $2\Upsilon = \dfrac{1}{R_i}\chi_{Ti}^2$. The condition of stationarity of the thermodynamic action $\mathscr{A}_2$ is

$$\delta\mathscr{A}_2 = 0: \quad \{\mathrm{d}_t \partial_{\dot{\xi}}[\phi + \delta^2\pi] - \partial_\xi \Upsilon\} \cdot \delta\xi = 0 \tag{6.5.12}$$

which gives the equations of motion

$$\ddot{\xi}_i - \beta_i^2 \xi_i = \frac{1}{R_i}[\dot{\chi}_{Ei} - \beta_i \chi_{Ei}] \tag{6.5.13}$$

in the diagonal representation.

The solutions of the equations of motion (6.5.13) are

$$\xi_i = c_i' \mathrm{e}^{-\beta_i t} + c_i'' \mathrm{e}^{\beta_i t} + \frac{1}{R_i} \mathrm{e}^{-\beta_i t} \int_0^t \mathrm{e}^{2\beta_i \tau} f_i(\tau)\, \mathrm{d}\tau \tag{6.5.14}$$

where c_i' and c_i'' are constants, determined by the initial conditions and

$$f_i(\tau) = \int_0^\tau \mathrm{e}^{-\beta_i \tau'} [\dot{\chi}_{Ei} - \beta_i \chi_{Ei}]\, \mathrm{d}\tau' \tag{6.5.15}$$

Now, if we integrate the first term by parts, we get

$$f_i(\tau) = \mathrm{e}^{-\beta_i \tau} \chi_{Ei} \tag{6.5.16}$$

This leads us to conclude that the original set of second order, inhomogeneous linear differential equations (6.5.13) can be written as two sets of first order homogeneous linear differential equations

$$(R_i \mathrm{d}_t + S_i)\xi_i = \chi_{Ei}(t) \tag{6.5.17}$$

$$(R_i \mathrm{d}_t - S_i)\xi_i = \chi_{Ei}(t) + h_i(t) \tag{6.5.18}$$

where

$$h_i(t) = -2\beta_i \mathrm{e}^{-\beta_i t} \int_0^t \mathrm{e}^{\beta \tau} \chi_{Ei}(\tau)\, \mathrm{d}\tau \tag{6.5.19}$$

We immediately recognise the first set of equations (6.5.17) as the half-degree of freedom variational equations (5.2.21). In the same way as before, the thermodynamic variational principle distinguishes between the two sets of variational equations (6.5.17) and (6.5.18). The thermodynamic action (6.5.10) can be written as

$$\mathscr{A}_2 = \frac{1}{2R_i} \int_{t_1}^{t_2} (R_i \dot{\xi}_i + S_i \xi_i - \chi_{Ei})^2\, \mathrm{d}t \tag{6.5.20}$$

Evaluating (6.5.20) with the solutions of the variational equations (6.5.17), which are the particular solutions of (6.5.14) with $c_i'' = 0$, we again obtain (6.5.8).

Thus, the nonlinear thermodynamic process will follow the 'thermodynamic' path that is described by (6.5.17) and not the 'anti-thermodynamic' path that is given by (6.5.18) (Lavenda, 1977). The thermodynamic variational principles of linear and nonlinear thermodynamics, (6.5.2) and (6.5. 9), respectively, can be interpreted as a measure of the error: in linear thermodynamics, the error is represented by the deviation of the dissipative force from the thermodynamic force. In nonlinear thermodynamics, it is represented by the deviation of the net internal force, $\chi_T - \chi_D$, from the external force. Here, we can observe the close connection between thermodynamic variational principles and the thermodynamic theory of fluctuations, where the deviations from the thermodynamic path are caused by a Langevin force (Onsager and Machlup, 1953; Lavenda, 1977).

Moreover, both the linear and nonlinear thermodynamic variational principles give the variational equations that describe the thermodynamic path when the convention of varying the velocities for a fixed configuration is imposed. This convention has a significance in terms of the theory of fluctuations: we are interested in calculating the change in the generalised displacement during a macroscopically small time interval. In other words, we are concerned with the determination of the path of an irreversible process which is described in terms of a finite difference equation. In the limit as the time interval is allowed to approach zero, we obtain the variational equations of the thermodynamic path.

It is a general property of the kinetic formulation of the thermodynamic variational principles that we always come out with two paths: the thermodynamic path and its mirror image. This is attributed to the fact that the thermodynamic actions are conservative in the classical mechanical sense. In other words, the thermodynamic Lagrangian, which is the sum of dissipation functions, possesses a symmetry with respect to past and future. The only distinction where past and future events is made is in the requirement that the thermodynamic action be a minimum. This involves the entropy explicitly and introduces the notion of irreversibility.

The kinetic formulation of the thermodynamic variational principles clearly distinguishes between the entropy which is a function of state, i.e. a function only of the endpoints of the path, and the dissipation functions which are functions of the path as well as the endpoints. This is to say that the entropy is a point function whereas the dissipation functions are not. In the language of fluctuation theory, we say that the dissipation functions describe the dynamic effects of correlations between two non-equilibrium states (Lavenda, 1977). The entropy production has a mute role in the kinetic formulation of the thermodynamic variational principles. We have already remarked that in the absence of the notion of irreversibility, there exists a symmetry with respect to past and future and this is manifested by the two sets of variational equations that are mirror images of one another. When the notion of irreversibility is introduced in the variational principle, by requiring that the thermodynamic action must be a minimum, we do, in fact, obtain the variational equations that describe the thermodynamic path.

When we consider the kinetic formulation of the thermodynamic variational principle that yields the variational equations (5.2.31), which are comprised of single degrees of freedom, only formal differences are encountered. The dissipation function ϕ remains invariant, whereas the generating function acquires an addition functional dependence. For simplicity, we treat the case of a vanishing external force variation. Then the generating function is a quadratic function of the thermodynamic force which is now a function of the accelerations as well as the generalised displacements. In place of the stationarity condition (6.5.3), we now get

$$\delta\mathscr{A} = 0: \quad \{\mathrm{d}_t^2 \partial_{\ddot{\xi}} \Upsilon - \mathrm{d}_t \partial_{\dot{\xi}} \phi + \partial_{\xi} \Upsilon\} \cdot \delta\xi = 0 \tag{6.5.21}$$

In comparison with the stationarity condition (6.5.3), we note that the roles of the dissipation function ϕ and the generating function Υ have been partially inverted (Machlup and Onsager, 1953). In (6.5.3), Υ is analogous to a potential energy which is a function of the state alone, while ϕ is a measure of its rate of change. Alternatively, in (6.5.21), ϕ is a function of state, since the velocities are included in the set of state variables of systems composed of single degrees of freedom and Υ depends on the rate of change of the state in time.

The stationarity conditions (6.5.21) give the equations of motion

$$M_{ij}^2\ddddot{\xi}_j+(2S_{ij}M_{ij}-R_{ij}^2)\ddot{\xi}_j+S_{ij}^2\xi_j = 0, \qquad i = 1, 2, \ldots, N \tag{6.5.22}$$

Again, we observe that the solutions of (6.5.22) are superpositions of the solutions to

$$M_{ij}\ddot{\xi}_j+R_{ij}\dot{\xi}_j+S_{ij}\xi_j = 0, \qquad i = 1, 2, \ldots, N \tag{6.5.23}$$

$$M_{ij}\ddot{\xi}_j-R_{ij}\dot{\xi}_j+S_{ij}\xi_j = 0, \qquad i = 1, 2, \ldots, N \tag{6.5.24}$$

which are the factors of (6.5.22). We obtain two classes of system trajectories or paths that are mirror images of one another. The set of variational equations that minimise the thermodynamic action are (6.5.23), which describe the thermodynamic path. Each of the classes of solutions will be comprised of two trajectories which are thermodynamically, as well as kinetically, indistinguishable. Complex solutions of (6.5.23), significant of rotational motion, are not outlawed by our thermodynamic treatment so long as the phenomenological coefficient symmetries are not destroyed. In this sense, the applicability of nonlinear thermodynamics depends on the level of representation. What would destroy the coefficient symmetries in terms of half-degree of freedom variational equations may leave the symmetries intact when the half-degree of freedom variational equations are converted into single degree of freedom variational equations. The next chapter discusses the stability criteria of nonlinear thermodynamics in terms of the symmetries of the phenomenological coefficients. In addition, there is offered a slight generalisation of the thermodynamic analysis by deriving a complex power equation which will permit the comparison of thermodynamic and kinetic stability criteria.

References

Casimir, H. B. G. (1945). On Onsager's principle of microscopic reversibility. *Rev. mod. Phys.*, **17,** 343–350.

Glansdorff, P. and Prigogine, I. (1964). On a general evolution criterion in macroscopic physics. *Physica,* **30,** 351–374.

Gyarmati, I. (1965*a*). *Zsurn. Fiz. Himii (Moscow) T.*, **39,** 1489.

Gyarmati, I. (1965*b*). *Acta. Chim. Hung.*, **43,** 353.

Gyarmati, I. (1965*c*). *Periodica Polytech.*, **9,** 205.

Gyarmati, I. (1966). *Acta. Chim. Hung.*, **47,** 63.

Gyarmati, I. (1970). *Non-Equilibrium Thermodynamics.* Springer-Verlag, Berlin.

Kirkaldy, J. S. (1964). *Can. J. Phys.*, **42,** 1447.

Lavenda, B. H. (1972). Concepts of stability and symmetry in irreversible thermodynamics I. *Found. Phys.*, **2,** 161–179.

Lavenda, B. H. (1974). Principles and representations of nonequilibrium thermodynamics. *Phys. Rev. A*, **9**, 929–943.

Lavenda, B. H. (1977). The path integral formulation of nonequilibrium statistical mechanics. *La Rivista del Nuovo Cimento*, **7**, 229–276.

Machlup, S. and Onsager, L. (1953). Fluctuations and irreversible processes. II. Systems with kinetic energy. *Phys. Rev.*, **91**, 1512–1515.

Ono, S. (1961). Variational principles in thermodynamics and statistical mechanics of irreversible processes, *Advan. Chem. Phys.*, **3**, 267–321.

Onsager, L. (1931). Reciprocal relations in irreversible processes I. *Phys. Rev.*, **37**, 405–426; Reciprocal relations in irreversible processes II. *Phys. Rev.*, **38**, 2265–2279.

Onsager, L. and Machlup, S. (1953). Fluctuations and irreversible processes. *Phys. Rev.*, **91**, 1505–1512.

Prigogine, I. (1954). *Bull. Acad. Roy. Belg. Cl. Sci.*, **31**, 600.

Prigogine, I. (1965). *Non-Equilibrium Thermodynamics, Variational Techniques, and Stability*. (Eds. Donnelly, R., Herman, R. and Prigogine I.), Chicago University Press, Chicago.

Rayleigh, Lord (J. W. Strutt) (1873). *Proc. Lond. math. Soc.*, **4**, 357.

Routh, E. J. (1877). *A Treatise on the Stability of a Given State of the Motion*. Macmillan, London.

Tisza, L. and Manning, I. (1957). Fluctuations and irreversible thermodynamics. *Phys. Rev.*, **105**, 1695–1705.

General references on variational principles

Lanczos, C. (1949). *The Variational Principles of Mechanics*. University of Toronto Press, Toronto.

Yourgrau, W. and Mandelstam, S. (1960). *Variational Principles in Dynamics and Quantum Theory*, 2 ed. Pitman, London.

7 Quasi-Thermodynamic Stability Theory

The problem of stability lies at the very root of nonlinear thermodynamic processes. Instabilities that cause dynamic transitions in open systems are responsible for the qualitative difference between linear and nonlinear thermodynamics. In linear thermodynamics, the symmetry and positive definite forms of the matrices of certain quadratic forms prohibit particular forms of system motion and guarantee the stability of the non-equilibrium process. In nonlinear thermodynamics, this may no longer be true and consequently the problem of stability arises. Furthermore, we expect the symmetries of the phenomenological coefficient matrices to be related to the type of motion that is executed by the non-equilibrium process.

Nonlinear thermodynamics is therefore concerned with the stability properties of non-equilibrium stationary states. Stability analysis is conventionally carried out by considering the motion in the immediate neighbourhood of the stationary state. We can already begin to appreciate the notion of stability: if all system trajectories approach the stationary state in the long time limit, then the stationary state is said to be 'asymptotically stable'. Instability is defined by the negation of this property. Since we know that kinetic stability theory provides necessary and sufficient conditions of stability for time independent states, the question naturally arises as to the purpose of a thermodynamic theory of stability. In other words, what can a thermodynamic theory of non-equilibrium stationary state stability add that is not already contained in the kinetic stability analysis? In this chapter an answer is given to this question.

In section 7.1 the classic approach to stability theory that was developed by Poincaré and Liapounov is reviewed. Since the method of variational equations provides necessary and sufficient conditions of stability, it will provide us with a framework for the evaluation of thermodynamic stability criteria. This implies that all thermodynamic criteria of stability must necessarily be compatible with the kinetic stability conditions. At this point we begin to appreciate the utility of a thermodynamic theory of stability in that it may provide general criteria of system stability which is not made evident through the integration of the variational equations.

The derivation of the thermodynamic stability criteria is begun in section 7.2 by generalising the power equation method of chapter 5. This results in the 'complex' power which separates the elements of the variational equations into two classes, according to their effect upon the stability of the motion. In the complex power method, the antisymmetric components of the phenomenological coefficient matrices are shown to be related to both the nature of the system motion and its stability. The significance of both the phenomenological symmetries and antisymmetries is discussed in section 7.3, in terms of the forces which they represent. In section 7.4 the system forces are related to the stability of the motion, and in section 7.5 the mechanical bases of phenomenological symmetry and antisymmetry are derived. The chapter concludes with the analysis of nonlinear irreversible processes. Through the application of the averaging techniques of nonlinear mechanics, it is shown in section 7.6 that the 'averaged' complex power analysis of nonlinear irreversible processes yields necessary and sufficient criteria of stability.

This chapter arrives at the conclusion that the thermodynamic stability theory itself does not yield necessary and sufficient conditions of stability in all cases. This is in direct contradiction to the claim made by the Glansdorff–Prigogine theory (cf. section 7.2). Rather, an approach is developed in which thermodynamic and kinetic concepts of stability are intertwined—hence the name 'quasi-thermodynamic' stability theory.

7.1 Elements of kinetic stability theory

The foundations of modern kinetic stability theory were developed in approximately the same period by Pioncaré (1892) and Liapounov (1892). Both Poincaré and Liapounov worked out a method to determine the stability in terms of the variational equations. In Liapounov's work, it is referred to as the 'first method'. The first method or the method of variational equations is concerned with problems of 'infinitesimal' stability or stability in the 'small'. This is to say that Liapounov's first method applies only to the stability analysis of small motions about a time independent state or a state of steady motion. Liapounov later supplemented his first method by what he called the 'second method'. The second method covers a wider domain surrounding the time independent state. In other words, Liapounov's second method provides conditions for stability or instability in the 'large', which consists of a finite domain surrounding the stationary state or the whole of the phase space.

Both the first and second methods analyse the stability properties of ordinary differential equations of the form

$$\dot{\alpha}_i = f_i(\alpha_1, \ldots, \alpha_N) \tag{7.1.1}$$

where the α_i are the internal state variables. In kinetic stability theory they are regarded as generalised coordinates without any special geometrical significance (Minorsky, 1969). It is assumed that the functions f_i are continuous and twice differentiable. Moreover, it is assumed that a stationary state exists

$$\alpha_i = \alpha_{i0} \tag{7.1.2}$$

where α_{i0} is referred to as the 'unperturbed' solution. This solution can be both time dependent (that is, a periodic solution) or time independent (that is, a stationary state). The question now arises as to the stability of the stationary state or steady motion which is governed by the autonomous set of equations (7.1.1).

In Liapounov's first method, we suppose that the state, whose stability we wish to determine, has been perturbed by some means so that all the generalised coordinates become

$$\alpha_i(t) = \alpha_{i0} + \xi_i(t) \tag{7.1.3}$$

The functions $\xi_i(t)$ are the perturbations and we suppose these functions to be small enough so that higher than first degree terms can be neglected in (7.1.3). Substituting (7.1.3) into the kinetic equations (7.1.1) and developing them about the unperturbed solution leads to

$$\dot{\xi}_i = (\partial_{\alpha_j} f_i)_0 \, \xi_j \tag{7.1.4}$$

where all derivatives are to be evaluated at the stationary state or the state of steady motion. This is implied by the subscript zero. The linear variational equations (7.1.4) govern the motion of the perturbation only in the case that the perturbation was initially small enough for its motion to be described by a linear set of differential equations. We can now appreciate the origin of the concept of stability: it requires the perturbation to be initially small and remain so in the course of time. The motion is called stable and this property determines the stability of the unperturbed solution (7.1.2). If the motion negates this criterion then the unperturbed solution is said to be unstable. The perturbation will grow in time and this is indicative of the fact that the system is evolving from the unstable stationary state.

Quantitative statements regarding stability in the small can be obtained from an analysis of the roots, Ω_i, of the characteristic equation of the system (7.1.4). There are three cases:

(1) If all roots have negative real parts, $\mathrm{Re}(\Omega) < 0$, the stationary state is asymptotically stable.

(2) If there is at least one root with $\mathrm{Re}(\Omega) > 0$, the stationary state is unstable.

(3) If there are no roots with $\mathrm{Re}(\Omega) > 0$ but at least one with $\mathrm{Re}(\Omega) = 0$, then the linear variational equations are insufficient to determine the stability.

Situation (3) is referred to as a 'critical' case (Minorsky, 1962) in which the stability is governed by the nonlinear terms that have been neglected in the linear variational equations (7.1.4). The linear variational equations show only that the non-equilibrium stationary state is metastable.

The analysis of the roots of the characteristic equation provides necessary and sufficient conditions of stability or instability of the stationary state. It says nothing concerning the motion in a wider region surrounding the stationary state.

Liapounov's second method is concerned with the problem of stability in the large. It applies to the stability of the motion at a finite distance from the stationary state. The second method approaches the stability problem from a completely different point of view than the method of variational equations. It transforms the nonlinear kinetic equations (7.1.1) into a form which shows, without integration, whether the trajectory of the motion approaches the stationary state.

The idea is to construct a 'Liapounov function' $\mathscr{D}(\alpha_i)$ which is continuous, together with its first derivatives, in a certain region R surrounding the stationary state. At the stationary state

$$\mathscr{D}(\alpha_{i0}) = 0 \tag{7.1.5}$$

and everywhere else in the region R, $\mathscr{D}(\alpha_i)$ is positive. Pictorially, $\mathscr{D}(\alpha_i)$ resembles a bowl-shaped surface in phase space whose deepest point coincides with the stationary state. Stability, in the sense of the second method, requires

$$\mathscr{D}(\alpha_i) \geqq 0 \tag{7.1.6}$$

throughout R, where the equality sign applies to the stationary state, and its 'Eulerian' derivative, or the derivative along the trajectory, satisfies the inequality

$$\dot{\mathscr{D}} = \partial_{\alpha_i}\mathscr{D}\dot{\alpha}_i = \partial_{\alpha_i}\mathscr{D}f_i = f\cdot\nabla\mathscr{D} < 0 \tag{7.1.7}$$

Liapounov's theorems can now be formulated as follows (Minorsky, 1969):

Theorem 1 If $\mathscr{D}(\alpha_i)$ exists in some neighbourhood R of the stationary state then the motion is stable.

Theorem 2 If $\mathscr{D}(\alpha_i)$ and $\dot{\mathscr{D}}(\alpha_i)$ are of opposite signs in the region R then the motion is asymptotically stable.

Theorem 3 If $\mathscr{D}(\alpha_i)$ exists and is continuous with continuous first derivatives in R, and if $\dot{\mathscr{D}}(\alpha_i)$ is positive definite along with $\mathscr{D}(\alpha_i)$, then the stationary state is unstable.

It is important to bear in mind that there are no definite rules for determining $\mathscr{D}(\alpha_i)$, nor is there any guarantee that once $\mathscr{D}(\alpha_i)$ has been determined it will be unique. Furthermore, Theorem 2 is sufficient but not necessary for stability.

It will be recalled from section 4.2 that generalised thermodynamics identifies the second variation of the specific entropy $\frac{1}{2}\delta^2\eta$ as a Liapounov function. Its Eulerian derivative, $\frac{1}{2}\delta^2\dot{\eta}$, is to be evaluated along the trajectory of the motion (Glansdorff, Nicolis and Prigogine, 1974)

$$\tfrac{1}{2}\delta^2\dot{\eta} = \partial_{\xi_i}\tfrac{1}{2}\delta^2\eta\,\dot{\xi}_i = \partial_{\xi_i}\tfrac{1}{2}\delta^2\eta\cdot\partial_{\alpha_j}f_i\xi_j > 0 \tag{7.1.8}$$

It will now be appreciated that the 'trajectory' of the motion is not given by the nonlinear kinetic equations (7.1.1), as in the case (7.1.7), but rather by the linear variational equations (7.1.4). It is therefore extremely misleading to call $\frac{1}{2}\delta^2\eta$ a Liapounov function, since it would imply that $\frac{1}{2}\delta^2\eta$ determines the stability in the large, which it is clearly incapable of doing. Furthermore, the condition

$$\tfrac{1}{2}\delta^2\eta \leqq 0 \tag{7.1.9}$$

is always postulated (i.e. the assumption of local equilibrium) and never proved. Finally, it is not always the case that there will exist a Liapounov function for a given set of nonlinear kinetic equations. Consequently, if $\frac{1}{2}\delta^2\eta$ is a Liapounov function in one case, it may not be in another case. The Glansdorff–Prigogine

(1971) association of the $\frac{1}{2}\delta^2\eta$ criterion, in one sense, with the universality of the second law, and in another sense, with a Liapounov function, leads to the conclusion that *every* non-equilibrium process possesses a Liapounov function, namely $\frac{1}{2}\delta^2\eta$. This is clearly not so! If this is not their intention, why then is there no criterion or word of caution to be found in the Glansdorff and Prigogine (1971) monograph that would determine the cases in which $\frac{1}{2}\delta^2\eta$ is or is not a Liapounov function?

However, it is clear that if appropriate thermodynamic criteria can be found, the question of stability or instability reduces to the application of *definite criteria* and avoids the not always possible task of integrating the variational system.

7.2 The complex power method

In linear thermodynamics the question of stability does not arise. The Onsager reciprocal relations guarantee the existence of non-equilibrium thermodynamic potentials whose definite forms are criteria of stability. Moreover, there is only one physically admissible solution to the possible nonlinear kinetic equations in linear thermodynamics. At equilibrium, this solution coincides with that obtained from the principle of detailed balancing at equilibrium. The uniqueness of solution implies that this solution must be stable in all physical systems.

The question of stability is meaningful only in nonlinear thermodynamics. Here, there is no guarantee that the Onsager reciprocal relations are valid and there may be other solutions to the kinetic equations in addition to that one which reduces to the conditions of detailed balancing at equilibrium. When there exists more than one solution to the kinetic equations, no solution can be stable globally, implying that nonlinear thermodynamic processes may actually become unstable.

Nonlinearity is a necessary ingredient for instability. However, in nonlinear thermodynamics, the effect of nonlinearity is only indirectly observed in the magnitudes and signs of the differential coefficients of the variational equations (7.1.4). Therefore, nonlinear thermodynamics is concerned only with 'infinitesimal' stability or stability in the 'small'. The influence of nonlinearity in governing the evolution of nonlinear irreversible processes can only be observed directly by a nonlinear analysis of the kinetic equations (cf. section 7.6). Bearing this distinction in mind, we develop an approach that establishes the connections with (*a*) the power equation method of chapter 5, and (*b*) the methods of nonlinear mechanics. Obviously, such an approach will require more than non-equilibrium thermodynamic principles, since it must take the kinetics of the process into consideration.

Quasi-thermodynamic stability criteria are derived in terms of the symmetries or antisymmetries, and singularities of the phenomenological coefficient matrices of the variational equations (5.2.21) or (5.2.31), depending on whether the non-equilibrium process is described by variational equations of half-degrees or single degrees of freedom. Since one class of system motion (i.e. gyroscopic motion) will not be observed in the half-degree of freedom description, the analysis will be performed on the single degree of freedom variational equations (5.2.31). The analysis can always be reduced to the half-degree of freedom variational equations by setting $\mathbf{M} = \mathbf{0}$.

The antisymmetric components of the phenomenological matrices **R** and **S** are significant of rotational processes. If we were to multiply the variational equations (5.2.31) by their conjugate velocities and sum them to form the excess power equation, it would be seen that the antisymmetric components of **R** and **S** vanish. However, the destruction of phenomenological symmetry implies that the system will respond to a finite frequency perturbation and hence we have to consider the case in which the variables are complex. In this case, if we multiply the variational equations (5.2.31) by their complex conjugate velocities and sum, we get

$$\dot{\xi}^+ \mathbf{M}\ddot{\xi} + \tfrac{1}{2}\dot{\xi}^+ (\mathbf{R} \pm \mathbf{R}^T)\dot{\xi} + \tfrac{1}{2}(\dot{\xi}^+ \mathbf{S}\xi \pm \xi^+ \mathbf{S}^T \dot{\xi}) = \pi' + i\pi'' \tag{7.2.1}$$

where ξ^+ is the complex conjugate row vector of the column vector ξ. In the complex power equation (7.2.1) we have assumed

$$\mathbf{M} = \mathbf{M}^T \tag{7.2.2}$$

since the antisymmetric part of **M**, if it existed, would have no physical significance. If the matrix **M** were not symmetric then the acceleration terms would not have a simple form in the variational equations. Henceforth, we shall always assume (7.2.2).

Equation (7.2.1) is an expression for the complex power which is defined as follows (Lavenda, 1972):

$$\pi = \pi' + i\pi'' = -\dot{\xi}^+ \cdot \chi_E \tag{7.2.3}$$

where

$$\pi' = -\tfrac{1}{2}(\dot{\xi}^+ \chi_E + \dot{\xi} \cdot \chi_E{}^+) = \tfrac{1}{2}(\pi + \pi^*) \tag{7.2.4}$$

$$i\pi'' = -\tfrac{1}{2}(\dot{\xi}^+ \chi_E - \dot{\xi} \cdot \chi_E{}^+) = \tfrac{1}{2}(\pi - \pi^*) \tag{7.2.5}$$

We can relate the real and imaginary parts of the complex power to the phenomenological coefficients by writing

$$\mathbf{S} = \mathbf{S}^+ + \mathbf{S}^-; \qquad S_{ij} = S_{ij}^+ + S_{ij}^- \tag{7.2.6}$$

where

$$\mathbf{S}^+ = \tfrac{1}{2}(\mathbf{S} + \mathbf{S}^T); \qquad S_{ij}^+ = \tfrac{1}{2}(S_{ij} + S_{ji}) \tag{7.2.7}$$

$$\mathbf{S}^- = \tfrac{1}{2}(\mathbf{S} - \mathbf{S}^T); \qquad S_{ij}^- = \tfrac{1}{2}(S_{ij} - S_{ji}) \tag{7.2.8}$$

Similar expressions are obtained for the phenomenological coefficient matrix **R**. The expressions for the real and imaginary parts of the complex power are then given by

$$\pi' = -\tfrac{1}{2}\{\mathrm{d}_t(M_{ij}\dot{\xi}_i^*\dot{\xi}_j + S_{ij}^+\xi_i^*\xi_j) + 2R_{ij}^+\dot{\xi}_i^*\dot{\xi}_j + S_{ij}^-(\dot{\xi}_i^*\xi_j - \xi_i^*\dot{\xi}_j)\} \tag{7.2.9}$$

$$\pi'' = \frac{i}{2}\{S_{ij}^+(\dot{\xi}_i^*\xi_j - \xi_i^*\dot{\xi}_j) + M_{ij}(\dot{\xi}_i^*\ddot{\xi}_j - \ddot{\xi}_i^*\dot{\xi}_j) + 2R_{ij}^-\dot{\xi}_i^*\dot{\xi}_j + \mathrm{d}_t(S_{ij}^-\xi_i^*\xi_j)\} \tag{7.2.10}$$

where ξ_i^* is the complex conjugate of ξ_i.

Prior to investigating the significance of the complex power expressions, let us compare it with the complex variable representation of the excess entropy production that has been given by Glansdorff and Prigogine (1971). They write the excess entropy production as

$$\delta^2\sigma_S = -\tfrac{1}{2}(\boldsymbol{\iota}^*\cdot\boldsymbol{\chi}+\boldsymbol{\iota}\cdot\boldsymbol{\chi}^*) \tag{4.2.30}$$

and define the so-called 'mixed' entropy production as

$$\sigma_S^M = -\frac{i}{2}(\boldsymbol{\iota}^*\cdot\boldsymbol{\chi}-\boldsymbol{\iota}\cdot\boldsymbol{\chi}^*) \tag{4.2.31}$$

where $\boldsymbol{\iota} = \dot{\xi}$ and $\boldsymbol{\chi}$ is the thermodynamic force variation. In terms of the phenomenological coefficients, (4.2.30) and (4.2.31) can be written explicitly as

$$\delta^2\sigma_S = -\tfrac{1}{2}\{S_{ij}^+(\dot{\xi}_i^*\xi_j+\xi_i^*\dot{\xi}_j)+S_{ij}^-(\dot{\xi}_i^*\xi_j-\xi_i^*\dot{\xi}_j)\} \tag{7.2.11}$$

$$\sigma_S^M = -\frac{i}{2}\{S_{ij}^+(\dot{\xi}_i^*\xi_j-\xi_i^*\dot{\xi}_j)\} \tag{7.2.12}$$

The Glansdorff–Prigogine criterion for the stability of a single normal mode is that (7.2.11) must be positive definite (Glansdorff and Prigogine, 1971). This will only be true in the case $\mathbf{S} = \mathbf{S}^T$ and consequently the second member vanishes identically. Thus, once there appears an antisymmetric part to the phenomenological matrix, we cannot indiscriminately apply the stability criteria of non-equilibrium thermodynamic potentials. Furthermore, it is clear that $\delta^2\sigma_S$ in (7.2.11) is *not* the excess entropy production since $\mathbf{S} \neq \mathbf{S}^T$.

In fact, the sign criterion that (7.2.11) must be positive definite in all stable processes does not come from the requirement that the excess entropy production must be positive definite in the Glansdorff–Prigogine theory. From equations (9.55) and (9.56) of their book, we see that the so-called stability criterion $\delta^2\sigma_S \geqq 0$ is derived from the fact that $\frac{1}{2}\delta^2\eta \leqq 0$. However, since σ_S is not the time derivative of $\frac{1}{2}\delta^2\eta$, due to the fact that there are antisymmetric components of the phenomenological coefficients in (7.2.11), the criterion

$$\tfrac{1}{2}\delta^2\eta \leqq 0 \Rightarrow \delta^2\sigma_S \geqq 0 \tag{7.2.13}$$

is completely fallacious. Let us now return to the expression for the complex power components, (7.2.9) and (7.2.10), and discuss their significance with respect to the necessary and sufficient conditions of stability laid down by Liapounov in his first method.

Since Liapounov's first method requires that there should be at least one root of the characteristic equation with a real positive part as the criterion for instability, cf. criterion (2), it suffices to consider a single mode, namely

$$\xi_i = \Xi_i\exp(\Omega t) \tag{7.2.14}$$

with

$$\Omega = \Omega_1 + i\Omega_2 \tag{7.2.15}$$

Quasi-thermodynamic criteria of stability can be obtained by placing the system along the imaginary axis in the complex frequency plane (cf. section 7.4). First, let us consider the case in which both the phenomenological coefficient matrices **R** and **S** are symmetric. Substituting (7.2.14) and its complex conjugate into expressions (7.2.9) and (7.2.10) leads to

$$\pi'(\Omega_2) = -\Omega_2^2 R_{ij}^+ \Xi_i^* \Xi_j = -2\phi \tag{7.2.16}$$

$$\pi''(\Omega_2) = -\tfrac{1}{2}\Omega_2(M_{ij}\Omega_2^2 - S_{ij}^+)\Xi_i^* \Xi_j \tag{7.2.17}$$

Although this is the case which is most frequently encountered in electric circuits (Guillemin, 1953), it has little relevance in nonlinear thermodynamics. It does, however, illustrate the difference between the real and imaginary components of the complex power. Expressions (7.2.16) and (7.2.17) describe small vibrations about an equilibrium position. The real part of the complex power $\pi'(\Omega_2)$ is related to the stability of the process; that is, the system must dissipate positively if it is to be stable. The imaginary part of the complex power $\pi''(\Omega_2)$ determines the natural frequencies of vibration from the condition

$$\pi''(\Omega_2) = 0 \tag{7.2.18}$$

In system analysis, $\pi''(\Omega_2)$ is a measure of the energy exchanges between the two forms of energy storage element. The general conclusion which can be drawn from these observations is that the terms appearing in $\pi'(\Omega_2)$ affect the stability of the process while those in $\pi''(\Omega_2)$ do not.

It is also possible to interpret (7.2.16) as the power required to compensate system dissipation. Under the action of an external force, the system will vibrate continuously about its equilibrium position. The motion is reversible and consequently the excess entropy production vanishes, as is clearly seen by comparing (7.2.9) and (7.2.16). From this we conclude that, in the presence of an external source of energy, the process may be dissipative while the motion is completely reversible! In chapter 5, we have stressed the duality which exists between the energy and entropy representations of the thermodynamic principle of the balance of power. In the energy representation, we would obtain the internal energy as the integral of the motion when the power compensates exactly the dissipated energy. This testifies to the fact that the system is effectively conservative and the motion is reversible. A mechanical example would be the case of forced oscillations.

We have already mentioned that this case is of little interest in nonlinear thermodynamics. It does, however, point out the significant difference between dissipation and irreversibility. More interesting is the case in which the nonlinear thermodynamic process is 'active'. Such a process is able to channel externally supplied energy into destabilising system modes. This is manifested by destruction of the symmetry of the phenomenological coefficient matrix, **S**.

As a second case, let us consider the coefficient matrices **R** and **S** to be non-symmetric. Again we place the system on the imaginary axis of the complex frequency plane and obtain the expression for the complex power as

$$\pi(\Omega_2) = \Omega_2\{\Omega_2 R_{ij} + i(M_{ij}\Omega_2^2 - S_{ij})\}\,\Xi_i^* \Xi_j \tag{7.2.19}$$

The real and imaginary parts of the complex power are obtained by introducing the complex amplitudes

$$\Xi_j = \Xi_j' - i\Xi_j'' \qquad \Xi_j',\ \Xi_j'' \quad \text{real} \tag{7.2.20}$$

and their complex conjugates into expression (7.2.19). The result is

$$\pi'(\Omega_2) = -\Omega_2\{\Omega_2 R^+_{ij}(\Xi'_i\Xi'_j + \Xi''_i\Xi''_j) - S^-_{ij}(\Xi'_i\Xi''_j - \Xi'_j\Xi''_i)\} \tag{7.2.21}$$

$$\begin{aligned}\pi''(\Omega_2) = -\frac{\Omega_2}{2}\{&(M_{ij}\Omega_2^2 - S^+_{ij})(\Xi'_i\Xi'_j + \Xi''_i\Xi''_j)\\ &-R^-_{ij}\Omega_2(\Xi'_i\Xi''_j - \Xi'_j\Xi''_i)\}\end{aligned} \tag{7.2.22}$$

We can no longer draw the same conclusions concerning the stability of the nonlinear thermodynamic process as we did in the previous case where both the phenomenological coefficient matrices were symmetric. However, we can conclude that the terms appearing in $\pi'(\Omega_2)$ will have a first order effect upon the stability of the motion, whereas those terms in $\pi''(\Omega_2)$ will not. We thus have a classification of the elements of the variational equations according to their effect on the stability of the non-equilibrium process. It is now important to relate the antisymmetric components of the phenomenological coefficients with the non-conservative system forces, of which they are manifestations. Moreover, we will be able to make some definite statements concerning the limits of stability from an analysis of the real part of the complex power that is based upon the results of Liapounov's first method. It will be appreciated that this is the antithesis of the Glansdorff–Prigogine theory which imposes non-existent thermodynamic criteria on the kinetic processes.

7.3 The significance of the antisymmetric components of the phenomenological coefficient matrices

We now inquire into the physical significance of the antisymmetric components of the phenomenological coefficient matrices **R** and **S**. Our investigation is facilitated greatly by the fact that although the variational equations (5.2.31) describe the evolution of internal state variables about a non-equilibrium stationary state, they could equally as well describe a mechanical process. Equations of the same form have the same solution so that we can rely upon the analogy with mechanical systems.

The phenomenological coefficients are related to the system forces through generalised equations of state (for example, Hooke's law). The presence of antisymmetric phenomenological coefficients implies that not all the system forces are conservative. In other words, the motion is no longer potential motion. This directs us to an analysis of the system forces rather than the non-equilibrium thermodynamic potentials. In contrast to linear thermodynamics, we can say that the roles of non-equilibrium thermodynamic potentials and system forces are interchanged in nonlinear thermodynamics (cf. chapter 8).

A classification of the system forces can be achieved on the basis of the functional dependencies of the forces. Forces which are velocity-dependent are referred to as 'motional' forces. There are two types: those which do negative work and those that do no work at all. These forces are referred to as 'dissipative' and 'gyroscopic' forces, respectively. All types of frictional force are included in the class of dissipative forces, while the Coriolis and Lorentz forces are examples of gyroscopic forces. The presence or absence of gyroscopic forces is frame-dependent. The presence of dissipative forces is manifested by the symmetric component of the phenomenological coefficient matrix **R**. The antisymmetric part of **R** is associated with gyroscopic forces. We already know from the components of the complex power, (7.2.21) and (7.2.22), that the two types of velocity-dependent force are distinguishable on the basis of their effect upon the stability of the motion.

Forces which are position-dependent are known as 'positional' forces. If the position-dependent force is conservative, being derivable from a scalar potential, then it is referred to as a 'non-circulatory' force. The name is borrowed from hydrodynamics and implies a specific form of 'flow' pattern (cf. section 8.2). In non-equilibrium thermodynamics the non-circulatory force is the thermodynamic force which is the negative gradient of the entropy production in the phase space of the internal state variables. The non-conservative component of the positional force is known as the 'circulatory' force. Unlike non-conservative motional forces, a finite rate of working is associated with circulatory forces. The presence of circulatory forces is indicated by the antisymmetric part of the phenomenological coefficient matrix **S**. Consulting the expression for $\pi'(\Omega_2)$, (7.2.21), we know that circulatory forces have a first order effect upon the stability of non-equilibrium stationary states. As we shall see from our field analysis of chapter 8, circulatory forces are responsible for rotational motion about non-equilibrium stationary states. In mechanics, circulatory forces have been known for a long time. Thomson and Tait (1886) describe processes that contain circulatory forces as those in which 'energy without limit may be drawn . . . by guiding it perpetually through a returning cycle of configurations'.

7.4 Stability of non-equilibrium stationary states

Our aim is to obtain definite criteria for stability and instability in terms of the complex power expression (7.2.19). There is only one way to do so and that is by deriving all the quasi-thermodynamic stability criteria from Liapounov's first method. The kinetic stability criteria are derived by considering the homogeneous set of variational equations (5.2.31) with $\chi_{Ei} = 0$. Introducing the solutions (7.2.14) into the homogeneous set of variational equations reduces them to algebraic equations which will have non-trivial solutions if and only if

$$|M_{ij}\Omega^2 + R_{ij}\Omega + S_{ij}| = 0 \tag{7.4.1}$$

Equation (7.4.1) is referred to as the 'characteristic' or 'secular' equation. According to Liapounov's first method, stability or instability is determined by the real parts of the roots of (7.4.1). These criteria can find expression in the complex power analysis, as we shall now show.

We begin our analysis by considering both phenomenological coefficient matrices to be symmetric. The stability criteria, obtained from Liapounov's first method, are succinctly expressed as

$$|R_{ij}| \geq 0 \qquad \text{and} \qquad |S_{ij}| \geq 0 \tag{7.4.2}$$

The equality signs apply to the critical cases which we have discussed in section

7.1. If $|R_{ij}| = 0$ then there will be at least one pair of complex conjugate roots with a zero real part and a non-vanishing imaginary part. In the complex frequency plane, the state of metastability is achieved at a finite frequency. Alternatively, if $|S_{ij}| = 0$, there is at least one root whose real part is zero together with its imaginary part. The first critical case corresponds to undamped oscillations about a non-equilibrium stationary state, whereas the second critical case determines the 'bifurcation' point (or fork) where two stationary solutions of the kinetic equations coalesce (cf. section 8.1). In other words, if $|S_{ij}| = 0$ then the system admits stationary solutions in addition to the trivial one. These two critical cases are indicative of nonlinear, dynamic cooperative phenomena which are discussed in the next chapter. Here, only the criteria of non-equilibrium stationary state stability are considered.

According to the complex power method, we place the system along the imaginary axis of the complex frequency plane. This state will not be metastable, since there is a non-vanishing external force. We can consider the action of the external force as a perturbation of an otherwise metastable state. It is for this reason that we have set the real part of the frequency equal to zero. The real part of the complex power is given by (7.2.16). According to the first stability criterion in (7.4.2), we have

$$\pi'(\Omega_2) \leq 0 \tag{7.4.3}$$

in all stable processes. The equality sign applies to the first critical case. The natural frequencies are determined by setting the imaginary part of the power equal to zero. Since the matrices are symmetric, there exists an orthogonal matrix which reduces simultaneously **M** and **S** to diagonal form. The imaginary part of the complex power (7.2.17) can therefore be written as

$$\pi''(\Omega_2) = -\frac{\Omega_2}{2}(M_k\Omega_2^2 - S_k)|\hat{\Xi}_k|^2 \tag{7.4.4}$$

and

$$\hat{\xi}_k = \hat{\Xi}_k \exp(i\Omega_2 t) \tag{7.4.5}$$

are the 'normal' coordinates. Setting (7.4.4) equal to zero, we obtain

$$\Omega_2^{(k)} = \pm(S_k/M_k)^{1\ 2} \tag{7.4.6}$$

which are the natural frequencies of vibration. The second criterion of stability in (7.4.2) requires all the natural frequencies of vibration to be real. From (7.4.6) we observe that the second critical case occurs when a natural frequency vanishes.

In terms of non-equilibrium thermodynamic potentials, the first criterion in (7.4.2) implies that the dissipation function is a positive semidefinite form, whereas the second criterion establishes the fact that the second variation of the entropy is a negative semidefinite form. These remarks give rise to two theorems of stability:

Theorem I If all system forces are conservative and if $-\frac{1}{2}\delta^2\eta$ and ϕ are both positive definite then the stationary state is stable.

Theorem II If all system forces are conservative and if either $\frac{1}{2}\delta^2\eta$ or ϕ vanish then the stability is critical.

Critical stability is defined in the Liapounov sense: there is at least one root of the characteristic equation (7.4.1) with a vanishing real part. However, stability in the Lagrange sense considers only one critical case where $\frac{1}{2}\delta^2\eta = 0$ (Ziegler, 1968). In other words, stability in the Lagrange sense requires the conditions of the following theorem of Lagrange: if all system forces are conservative and if $\frac{1}{2}\delta^2\eta$ is negative definite then the stationary state is stable.

In the case that there is a non-conservative positional force, the complex power method cannot be used to determine the stability of the motion. As we have already remarked, this is in direct contradiction to the universal stability criteria of Glansdorff and Prigogine (1971). In fact, in the case of nonlinear thermodynamic processes that are comprised of half-degrees of freedom ($\mathbf{M} = \mathbf{0}$ and $\mathbf{R} = \mathbf{R}^T$), we have

$$\pi'' = -\sigma_S^M \tag{7.4.7}$$

and it behoves us to find a criterion from kinetic stability analysis which will provide a sign criterion for (7.4.7) that would supposedly determine the direction of rotation about a non-equilibrium stationary state. Rather, we have to content ourselves with the distinction between the two types of non-conservative forces in respect to their influence on the stability of the motion.

From the fact that the antisymmetric part of $\mathbf{R}$ appears only in the imaginary part of the complex power, we conclude that gyroscopic forces do not have a first order effect upon stationary state stability. Consider the case in which

$$\mathbf{R} = -\mathbf{R}^T \tag{7.4.8}$$

The characteristic equation (7.4.1) reduces to

$$|M_{ij}\Omega^2 + R_{ij}^-\Omega + S_{ij}^+| = 0 \tag{7.4.9}$$

Interchanging rows and columns of the determinant leaves it unchanged. Because $\mathbf{M} = \mathbf{M}^T$, $\mathbf{S} = \mathbf{S}^T$, and (7.4.8), the interchange of terms in the determinant merely converts Ω into $-\Omega$. The roots of the characteristic equation (7.4.9) appear in pairs that are equal and of opposite sign. Stability must therefore be considered in the Lagrange sense: the roots Ω must all be real and negative. Equivalently, we could require the square of the natural frequencies (7.4.6) to be all real and positive. This condition is expressed by the following theorem:

Theorem III A nonlinear thermodynamic process that is comprised of conservative forces cannot be made unstable by gyroscopic forces.

Gyroscopic forces may, however, have a more subtle effect upon system stability. In the absence of dissipativity and in the case where the second variation of the entropy is not negative definite, the expression for the imaginary part of the power (7.2.22) indicates that gyroscopic forces may have a stabilising influence on the motion. However, in the presence of dissipation, gyroscopic forces are unable to convert unstable motion into stable motion; rather, they can only lengthen the time of evolution away from an unstable configuration (for example, a sleeping top). This anomalous effect was first studied in mechanical systems by Thomson and Tait (1886).

Circulatory forces, unlike gyroscopic forces, have a very definite effect upon the stability of the motion. In the complex power analysis, their presence is observed in the last set of terms in the expression for the real part of the power (7.2.21). We recall that in the Glansdorff–Prigogine theory, the circulatory forces enter into the expression for the excess entropy production (7.2.11). On account of these non-conservative forces, we are no longer able to apply the stability criterion (7.4.3). It is surprising that Glansdorff and Prigogine claim

$$\delta^2 \sigma_S \geqq 0 \tag{7.4.10}$$

to be universally valid! All that we can infer in the presence of circulatory forces is summarised in the following theorem:

Theorem IV A nonlinear thermodynamic process that is comprised of conservative forces can be stabilised or destabilised by circulatory forces.

The effect that the non-conservative positional forces will have upon stability cannot be determined by the complex power method. We must necessarily resort to Liapounov's first method in order to determine whether the circulatory forces stabilise or destabilise the motion.

We emphasise the fact that the analysis of circulatory forces in nonlinear thermodynamics is by no means academic. As we shall see in the next chapter, circulatory forces are the cause of rotations about non-equilibrium stationary states. The first mention of circulatory forces is found in Thomson and Tait (1886). They thought the appearance of circulatory forces to be very improbable, since the only way to destroy the conservative nature of the system motion would be to connect the system with an external energy source. Yet, this is precisely the situation of non-equilibrium processes that occur in open systems. The ability of nonlinear thermodynamic processes to use the energy supplied by reservoirs for particular forms of system motion is at the very heart of dynamic, cooperative phenomena. Exemplary of such a process is the Bénard instability in hydrodynamics. However, the analysis of the effects due to non-conservative positional forces do not fall within the limits of classical non-equilibrium thermodynamics. It requires a new concept of a 'force field', which is introduced in the next chapter.

7.5 The mechanical bases of phenomenological symmetries and antisymmetries

We recall from section 2.1 that the reciprocal relations between equivalent causes and effects had been known for almost a century prior to Onsager. We now discuss, in somewhat greater detail, the mechanical basis for the class of reciprocal relations that bears Onsager's name.

Probably the most convincing mechanical explanation of the reciprocal relations, including those of the dissipative forces, is to be found in Rayleigh (1877). He performed his analysis in terms of small vibrations about a point of equilibrium. The variational equations (5.2.31) are written in the form

$$\Theta_{ij}\xi_j = \chi_i, \qquad i = 2, \ldots, N \tag{7.5.1}$$

where Θ_{ij} is the quadratic operator

$$\Theta_{ij} = M_{ij}D^2 + R_{ij}^{+}D + S_{ij}^{+}, \qquad D = d_t \tag{7.5.2}$$

In (7.5.1) we have dropped the subscript on the external force, since no confusion can arise.

Since all coefficient matrices are assumed to be symmetric, it follows that

$$\Theta_{ij} = \Theta_{ji} \tag{7.5.3}$$

Provided that

$$\Delta = |\Theta_{ij}| \neq 0 \tag{7.5.4}$$

we may use Cramer's rule to express the generalised displacements as linear combinations of the external forces, namely

$$\xi_i = \frac{\Gamma_{ji}}{\Delta}\chi_j, \qquad i = 1, 2, \ldots, N \tag{7.5.5}$$

where Γ_{ji} is the cofactor of Θ_{ji}. If all but two external forces vanisn, then the set of equations (7.5.5) reduces to

$$\Delta\xi_1 = \Gamma_{11}\chi_1 + \Gamma_{21}\chi_2 \tag{7.5.6}$$

$$\Delta\xi_2 = \Gamma_{12}\chi_1 + \Gamma_{22}\chi_2 \tag{7.5.7}$$

Now consider two cases of the motion, the first for which $\chi_2 = 0$ and the second for which (using primes) $\chi_1' = 0$. In the first case, (7.5.7) becomes

$$\xi_2 = (\Gamma_{12}/\Delta)\chi_1; \qquad \chi_2 = 0 \tag{7.5.8}$$

while in the second case (7.5.6) reduces to

$$\xi_1' = (\Gamma_{21}/\Delta)\chi_2'; \qquad \chi_1' = 0 \tag{7.5.9}$$

Since (7.5.4) is a symmetrical determinant, we have the reciprocity relation

$$\Gamma_{ij} = \Gamma_{ji} \tag{7.5.10}$$

from which it follows

$$\chi_1 : \xi_2 = \chi_2' : \xi_1' \tag{7.5.11}$$

Expressed in words, (7.5.11) states that if a force χ_1 were to act alone in the system, it would cause a response ξ_2 that would be equal to the response ξ_1' were the force χ_2' to act alone. Notice that we have used the reciprocal relation to give them a mechanical explanation.

The symmetry relation (7.5.10) constitutes a single class of reciprocity laws to which the Onsager relations belong. Other classes of reciprocal relations are known, such as those between coupled transport processes For this class of reciprocal relations, attempts were made to derive them trom known principles of classical mechanics. Maxwell (1860) argued that the symmetry relations between the phenomenological coefficients of coupled diffusional processes are a consequence of Newton's third law of 'equal and opposite reactions'. Specifically, the force exerted by a species A on species B is equal in magnitude and opposite in direction to that exerted by B on A (Johnson, 1951). Truesdell (1962) has objected to this argument on two counts: (1) the diffusion force is not the entire force of one constituent on another, and (2) the force of diffusion is not a *binary* force, so that Newton's third law cannot be applied to the case of more than two constituents. However, we must be careful not to confuse this class of reciprocal relations with the Onsager relations (cf. Truesdell, 1962), since the 'flux' is not the time rate of change of an extensive variable.

In an analogous way that a mechanical explanation can be given to phenomenological symmetry, we expect there to be a mechanical basis for phenomenological antisymmetry. The analysis of Rayleigh (1877) can be presented as an equivalence of different sets of work functions. That is, the phenomenological symmetry relation (7.5.10) establishes the following equivalence among the work functions:

$$\chi_i \xi'_i = \chi'_i \xi_i \tag{7.5.12}$$

It can easily be shown that in the case of two sets of equivalent forces and generalised displacements, the conditions

$$\chi_2 = 0 \qquad \text{and} \qquad \chi'_1 = 0 \tag{7.5.13}$$

yield the force–displacement ratios (7.5.11). The mechanical significance of this is quite clear. If we now want to apply the same type of reasoning to the case of phenomenological antisymmetry we are immediately confronted by the fact that there are no generalised work functions for the non-conservative forces. Another approach is called for that is based on the properties of Pfaffian forms.

Consider two types of independent variation, Δ and δ. The variations in the corresponding work functions are

$$\omega_\delta = \chi_i \delta\xi_i \tag{7.5.14}$$

$$\omega_\Delta = \chi_i \Delta\xi_i \tag{7.5.15}$$

Operating on (7.5.14) and (7.5.15) by Δ and δ, respectively, and then taking the difference of these two expressions, leads to

$$\Delta\omega_\delta - \delta\omega_\Delta = \Delta(\chi_i \delta\xi_i) - \delta(\chi_i \Delta\xi_i) \tag{7.5.16}$$

Now, since δ and Δ commute, (7.5.16) can be written as

$$\Delta\omega_\delta - \delta\omega_\Delta = (\Delta\chi_i \delta\xi_i - \delta\chi_i \Delta\xi_i) = 2Q_{\bar{ij}}\delta\xi_i \Delta\xi_j \tag{7.5.17}$$

where

$$Q_{\bar{ij}} = \tfrac{1}{2}(\partial_{\xi_j}\chi_i - \partial_{\xi_i}\chi_j) \tag{7.5.18}$$

In the case that (7.5.18) vanishes, then $d\omega$ is an exact or perfect differential of some scalar potential. Alternatively, in the case that (7.5.18) does not vanish then the forces χ_i are non-conservative, since they cannot be derived from a scalar potential. It is the latter case which we will be concerned with in establishing the mechanical basis for phenomenological antisymmetry.

Let $\{\tilde{\xi}_i\}$ be a new set of generalised displacements that are derived from $\{\xi_i\}$ by some transformation which we need not specify. The antisymmetric phenomenological coefficients are defined in this basis as

$$\tilde{Q}_{\bar{ij}} = \tfrac{1}{2}(\partial_{\tilde{\xi}_j}\tilde{\chi}_i - \partial_{\tilde{\xi}_i}\tilde{\chi}_j) \tag{7.5.19}$$

Since the expression $(\Delta\omega_{\delta}-\delta\omega_{\Delta})$ has obviously the same value in whatever basis we choose, it follows that

$$Q_{\overline{ij}}\delta\xi_i\Delta\xi_j=\tilde{Q}_{\overline{ij}}\delta\tilde{\xi}_i\Delta\tilde{\xi}_j \tag{7.5.20}$$

In classical mechanics, the left-hand side of (7.5.20) is called the 'bilinear covariant' of the Pfaffian differential form (7.5.14) (Whittaker, 1944). Since it can be expressed in the form (7.5.17) we have the following equivalence between the two sets of variables

$$\Delta\chi_i\delta\xi_i-\delta\chi_i\Delta\xi_i=\Delta\tilde{\chi}_i\delta\tilde{\xi}_i-\delta\tilde{\chi}_i\Delta\tilde{\xi}_i \tag{7.5.21}$$

Suppose that the variations Δ and δ refer to increments between any given orbit and an adjacent orbit. In other words, δ refers to that change which occurs when the system passes into an orbit given by

$$(\tilde{\xi}_1,\ldots,\tilde{\xi}_N,\tilde{\chi}_1,\ldots,\tilde{\chi}_i+\delta\tilde{\chi}_i,\ldots,\tilde{\chi}_N) \tag{7.5.22}$$

at time t_0, whereas Δ denotes the change when the system passes into the orbit

$$(\xi_1,\ldots,\xi_N,\chi_1,\ldots,\chi_j+\Delta\chi_j,\ldots,\chi_N) \tag{7.5.23}$$

at a different time t_1. In this interpretation, (7.5.21) reduces to

$$\Delta\chi_j:\Delta\tilde{\xi}_i=-\delta\tilde{\chi}_i:\delta\xi_j \tag{7.5.24}$$

Expressed in words: a change in $\tilde{\chi}_i$ causes a change in ξ_j that is equal to the change in $\tilde{\xi}_i$, with a *reversed* sign, which is caused by a change in χ_j, equal to that of the change in $\tilde{\chi}_i$, when all other forces and displacements are held constant. This is expressed in terms of the phenomenological coefficients as

$$Q_{ij}=-Q_{ji} \tag{7.5.25}$$

Expression (7.5.24) forms the basis of Helmholtz's reciprocal theorem. In physical terms, (7.5.24) states that a force of the first type causes a change in the displacement of the second type in the direction of the motion that is equal to the corresponding change in the displacement of the first type caused by a force of the second type when the motion is reversed. By reversed motion, we mean that which starts from a given position where each of the velocities of the particles change in sign, so that the system retraces its previous path only now passing through all states in the reverse order. The type of motion is guaranteed by the principle of 'dynamical reversibility' (Tolman, 1938) which can be used to derive the anti-reciprocal relations (7.5.24).

More can be said about the Pfaffian form (7.5.14). Using Stokes's theorem, we integrate (7.5.17) over an open surface S that is bounded by the closed curve C. We then obtain

$$\int_S(\partial_{\xi_j}\chi_i-\partial_{\xi_j}\chi_i)\Delta\xi_j\delta\xi_i=\oint_C\chi_i\delta\xi_i \tag{7.5.26}$$

The non-vanishing of the right-hand side of (7.5.26) testifies to the fact that the work function is multivalued. We will come across multivalued non-equilibrium thermodynamic potentials in our hydrodynamic analysis of rotational processes in section 8.2. Expression (7.5.26) is known as the 'circulation', and if the 'flow' is area preserving then the circulation is constant, namely

$$\mathrm{d}_t\oint_C\chi_i\delta\xi_i=0 \tag{7.5.27}$$

This is known as Kelvin's theorem in hydrodynamics. According to (7.5.27), the right-hand side of (7.5.26) is a relative integral invariant of the motion. The left-hand side of (7.5.26) is then an absolute integral invariant of order two.

It will now be appreciated that there exists a geometrical distinction between phenomenological symmetry and antisymmetry. The reciprocal relation (7.5.10) guarantees the existence of non-equilibrium thermodynamic potentials, namely the entropy production and dissipation function. These reciprocal relations can be referred to as the generalised Maxwell relations, since they reduce to the equilibrium Maxwell relation when $\mathbf{M} = \mathbf{R} = \mathbf{0}$. If non-equilibrium thermodynamic potentials exist then they are necessarily point functions. This is to say that a change in the potential, due to a transition from one non-equilibrium state to another, is equal to the difference between the values of the potential in the final and initial states. The basic geometrical element in the phase space is the 'distance'. On the other hand, phenomenological antisymmetry implies that the thermodynamic potentials are path dependent. Each time the system performs a closed cycle in phase space, the potential increases by a constant amount. The basic geometrical element in the phase space is now the 'area'. The mode of system motion is rotational and, as such, it does not admit a description in terms of scalar potentials. It requires a new approach in which the concept of a 'field' is employed (cf. chapter 8).

Thus far, we have shown how the power equation method, developed in chapter 5, can be generalised so as to include the actions of non-conservative forces. The complex power method is not purely thermodynamical in character, since it relies on the kinetic stability criteria of Liapounov's first method. It is now interesting to show how the complex power method can be generalised still further so as to include non-equilibrium processes that are truly nonlinear. In a similar way that the complex power method relies on kinetic stability theory, its nonlinear generalisation employs averaging techniques of nonlinear mechanics (Lavenda, 1973).

7.6 Stability of nonlinear irreversible processes

Up to this point, the complex power method has been used to determine the stability of the motion of nonlinear thermodynamic processes that have been displaced infinitesimally from a non-equilibrium stationary state. For system motion that enters into the nonlinear region surrounding a non-equilibrium stationary state, the method of variational equations and consequently the complex power method, hitherto developed, are inapplicable. The motion is no longer governed by linear differential equations.

One possible approach would be to find the non-equilibrium thermodynamic potential analogues of the Liapounov functions for each particular process. This appears to be neither a general nor fruitful approach, since certain classes of nonlinear process do not admit a Liapounov function type of analysis. Another approach would be to generalise the complex power method through the use of nonlinear averaging techniques (Lavenda, 1973). If we accept such an approach then we necessarily have to relinquish an instantaneous description of the motion. In general, nonlinear analyses are not concerned with an instantaneous description of the motion but rather they seek to determine invariant or 'fixed' points through which the system passes continuously at constant intervals of time (Minorsky, 1947; Minorsky, 1962). Hence, in contrast to linear stability analysis, we are now concerned with the 'average' power of nonlinear irreversible processes.

It will be recalled from sections 7.2 and 7.4 that the complex power is comprised of two components: a real part that is related to system stability through the criterion (7.4.3) and an imaginary part which is related to the phase lag between the applied external force and the rate of change of its conjugate extensive variable. One way to account for the phase lag is by defining a complex force (Lavenda, 1973)

$$-\chi_E = \chi_D + i\chi_T \tag{7.6.1}$$

and form the complex power according to (7.2.3), namely

$$\pi = \pi' + i\pi'' = (\chi_D + i\chi_T)\cdot\dot{\xi}^+ \tag{7.6.2}$$

If we now average the complex power over a characteristic period of the motion, τ, then we obtain an expression for the averaged power, namely

$$\bar{\pi} = \frac{1}{\tau}\int_0^\tau (\chi_D + i\chi_T)\cdot\dot{\xi}^+ \, dt \tag{7.6.3}$$

where the bar denotes the averaged value of the function. By defining an averaged power, we have shifted the emphasis from the original set of variables to the variables that describe the long time evolution of the nonlinear process. These variables are the amplitude A and phase ϕ which are assumed to be slowly varying functions of time over a scale which is much longer than that of a single period of the motion.

In the case of a single degree of freedom, we have

$$\xi = A \exp i\phi; \qquad \phi = \Omega t + \alpha \tag{7.6.4}$$

and the condition of quasi-periodicity over a single period of the motion is

$$\dot{\xi} = i\Omega A \exp i\phi \tag{7.6.5}$$

In order for (7.6.5) to be valid, we require

$$\dot{A} \ll \Omega A \tag{7.6.6}$$

which means that the build-up or decay of the nonlinear periodic phenomena is much slower than the quasi-periodic motion itself. Introducing (7.6.4) and (7.6.5) into (7.6.3) yields

$$\bar{\pi}(\Omega, A) = \bar{\pi}'(\Omega, A) + i\bar{\pi}''(\Omega, A) \tag{7.6.7}$$

which is analogous to the so-called 'describing' function of system theory (Blaquière, 1966).

With the aid of the averaged power expression (7.6.7), we can derive some interesting and intuitive stability criteria of states of steady motion that occur at a finite distance from a non-equilibrium stationary state. The states of steady motion or what we have previously referred to as 'fixed' points are determined from the condition

$$\overline{\pi}(\Omega_0, A_0) = 0 \tag{7.6.8}$$

in the absence of external forcing functions. The stationary amplitude A_0 is determined by the condition that the real part of the averaged power vanishes

$$\overline{\pi}'(\Omega_0, A_0) = 0 \tag{7.6.9}$$

while the stationary value of the frequency Ω_0 is obtained from

$$\overline{\pi}''(\Omega_0, A_0) = 0 \tag{7.6.10}$$

The state of steady motion can be a 'limit' cycle which is a periodic solution of the nonlinear kinetic equations. A characteristic difference between linear and nonlinear irreversible processes is that the amplitude is no longer determined by arbitrary initial conditions in the nonlinear case. Rather, it is determined by the differential equation itself! If the limit cycle is stable, then beginning at an arbitrary initial position the system will ultimately approach the nonlinear state of periodic motion and hence the name 'limit' cycle. We are now confronted with the problem of determining the stability of the state of steady motion.

In order to determine the stability characteristics, we perturb the state of steady motion. The amplitude and frequency become

$$A = A_0 + \delta A \tag{7.6.11}$$

$$\Omega = \Omega_0 + \delta\Omega_1 + i\delta\Omega_2, \qquad \delta\Omega_1,\ \delta\Omega_2 \text{ real} \tag{7.6.12}$$

The criterion of stability is obtained by observing that the variation of the amplitude in time is (Blaquière, 1966)

$$[A_0 + \delta A(t)]\exp(-\delta\Omega_2 t) \tag{7.6.13}$$

in the immediate neighbourhood of the state of steady motion. From the form of (7.6.13), it is apparent that the state of steady motion will be stable provided

$$\operatorname{sgn}[\delta A(t)] = \operatorname{sgn}(\delta\Omega_2) \tag{7.6.14}$$

for only then will the variation in the amplitude decay in time. The explicit stability condition is obtained by substituting (7.6.11) and (7.6.12) into the averaged power expression (7.6.7) and imposing the condition

$$\overline{\pi}(\Omega, A) = 0 \tag{7.6.15}$$

We then obtain

$$\overline{\pi}(\Omega_0 + \delta\Omega_1 + i\delta\Omega_2, A_0 + \delta A) = \overline{\pi}'(\Omega_0, A_0) + i\overline{\pi}''(\Omega_0, A_0) + (\partial_A\overline{\pi}' + i\partial_A\overline{\pi}'')_0\,\delta A + (\partial_\Omega\overline{\pi}' + i\partial_\Omega\overline{\pi}'')_0(\delta\Omega_1 + i\delta\Omega_2) + 0(\delta A) = 0 \tag{7.6.16}$$

where the partial derivatives are evaluated at the state of steady motion.

In order for the expansion (7.6.16), in terms of the perturbed variables, to be valid it is necessary that

$$\lim_{\delta A \to 0} 0(\delta A)/\delta A = 0 \tag{7.6.17}$$

In general, the frequency will be a function of the amplitude, being related by a type of dispersion equation. Then, if the deviation from the state of steady motion is not too great, the frequency will depend upon the amplitude in a linear fashion, say

$$\delta\Omega_1 = \lambda_1 \delta A \qquad \text{and} \qquad \delta\Omega_2 = \lambda_2 \delta A \tag{7.6.18}$$

with λ_1 and λ_2 constant. In view of the stability criterion (7.6.14), we conclude that $\lambda_2 > 0$ for the state of steady motion to be stable. The explicit expression for λ_2 is found by substituting (7.6.18) into (7.6.16). Dividing by δA we obtain

$$(\partial_A \bar{\pi}')_0 = -(\partial_\Omega \bar{\pi}')_0 \lambda_1 + (\partial_\Omega \bar{\pi}'')_0 \lambda_2 \tag{7.6.19}$$

$$(\partial_A \bar{\pi}'')_0 = -(\partial_\Omega \bar{\pi}'')_0 \lambda_1 - (\partial_\Omega \bar{\pi}')_0 \lambda_2 \tag{7.6.20}$$

since the first two terms in (7.6.16) vanish on account of the stationarity conditions (7.6.9) and (7.6.10). The pair of equations (7.6.19) and (7.6.20) can be solved for the coefficients appearing in (7.6.18). We then obtain the necessary and sufficient criterion of stability as

$$\lambda_2 > 0 : (\partial_A \bar{\pi}' \partial_\Omega \bar{\pi}'' - \partial_A \bar{\pi}'' \partial_\Omega \bar{\pi}')_0 > 0 \tag{7.6.21}$$

Blaquière (1966) has shown that the stability criterion of the state of steady motion (7.6.21) can be given an interesting geometrical interpretation in the $\bar{\pi}$ complex plane. The state of steady motion is located at the origin of the complex plane and at which point — the hodographs — the curves of equi-frequency Ω_0 and equi-amplitude A_0 intersect. In the $\bar{\pi}$ complex plane we can define two vectors

$$\boldsymbol{\beta}_1 = (\partial_A \bar{\pi}', \partial_A \bar{\pi}'')_0 \tag{7.6.22}$$

$$\boldsymbol{\beta}_2 = (\partial_\Omega \bar{\pi}', \partial_\Omega \bar{\pi}'')_0 \tag{7.6.23}$$

The stability criterion (7.6.21) can now be cast in the form of a vector product, namely

$$\boldsymbol{\beta}_1 \wedge \boldsymbol{\beta}_2 > 0 \tag{7.6.24}$$

Instability is signalled when the hodographs become parallel to one another.

A more detailed analysis of nonlinear irreversible processes would carry us too far afield from our main objectives. A truly nonlinear thermodynamic theory has yet to be developed. In this section some feasible approaches were offered, and section 9.7 will return to the problem in connection with nonlinear wave propagation.

References

Blaquière, A. (1966). *Nonlinear System Analysis*, Academic Press, New York.

Glansdorff, P., Nicolis, G. and Prigogine, I. (1974). The thermodynamic stability theory of non-equilibrium states. *Proc. natn. Acad. Sci. U.S.A.*, **71**, 197–199.

Glansdorff, P. and Prigogine, I. (1971). *Thermodynamic Theory of Structure, Stability, and Fluctuations*, Wiley–Interscience, London.

Guillemin. E. A. (1953). *Introductory Circuit Theory,* Wiley, New York.
Johnson, M. H. (1951). *Phys. Rev.,* **84,** 566.
Lavenda, B. H. (1972). Concepts of stability and symmetry in irreversible thermodynamics I. *Found. Phys.,* **2,** 161–179.
Lavenda, B. H. (1973). Thermodynamics of nonlinear, interacting irreversible processes II. *Found. Phys.,* **3,** 53–88.
Liapounov, A. M. (1892). Probleme générale de la stabilité du mouvement. *Ann. Fac. Sci. Toulouse.*
Maxwell, J. C. (1860). *Papers* **1,** 377.
Minorsky, N. (1947). *Introduction to Non-linear Mechanics,* J. W. Edwards, Ann Arbor, Michigan.
Minorsky, N. (1962). *Nonlinear Oscillations,* Van Nostrand, Princeton.
Minorsky, N. (1969). *Theory of Nonlinear Control Systems,* McGraw-Hill, New York.
Poincaré, H. (1892). *Les Méthodes Nouvelles de la Mécanique Céleste,* vol. 1, Gauthiers-Villars, Paris.
Rayleigh, Lord (J. W. Strutt) (1877). *Theory of Sound,* 1 ed., Macmillan, London.
Thomson, W. and Tait, P. G. (1886). *Treatise on Natural Philosophy,* Cambridge University Press, Cambridge.
Tolman, R. C. (1938). *The Principles of Statistical Mechanics,* Oxford University Press, London.
Truesdell, C. (1962). The mechanical basis of diffusion. *J. chem. Phys.,* **37,** 2336–2344.
Whittaker, E. T. (1944). *Analytical Dynamics,* Dover Pub., New York.
Ziegler, H. (1968). *Principles of Structural Stability,* Blaisdell, Waltham, Mass.

8 Field Thermodynamics

There is no continuity between linear and nonlinear thermodynamics processes. Linear thermodynamics treats, essentially, processes that relax monotonically to the state of equilibrium. Dynamic cooperative phenomena can only arise in nonlinear thermodynamics. If we limit ourselves to the analyses of spatially uniform processes then cooperative phenomena fall into two broad classes: multistationary state transitions and rotational processes. Among other things, these cooperative phenomena differ in respect to the number of degrees of freedom that are required to describe the non-equilibrium processes. Multistationary state transitions utilise only a half-degree of freedom, whereas rotational processes require at least a single degree of freedom.

We know that for a process of a half-degree of freedom there will always exist a scalar potential. We shall call this function a 'velocity' potential, since it is a potential function of the velocity field. The trouble with such an analysis occurs when the number of degrees of freedom of a non-equilibrium process increases. Non-equilibrium processes that display rotational motion destroy the exactness conditions which are the criterion for the existence of scalar potentials. In other words, for rotational motion, the non-equilibrium thermodynamic potentials become multivalued and we must search other ways to describe the non-equilibrium process. Analogous phenomena are found in hydrodynamics and electrodynamics, and we shall discover that there exists a wealth of similarities between rotational non-equilibrium processes and vortices in hydrodynamics and electric currents in electrodynamics. On the basis of these analogies, we shall be able to generalise nonlinear thermodynamics to include the dynamic phenomena which are most conveniently described in terms of velocity and force fields.

Multistationary state transitions represent a form of potential motion that implies freedom from rotation. The nature of the unstable transitions can be most easily characterised in terms of a scalar potential whose extremum property determines both the stability of the non-equilibrium stationary states and the direction of system motion (Lavenda, 1972*b*). That is, for multistationary state transitions, there exists a scalar function U whose negative gradient in the space of internal state variables $\{\alpha\}$ determines the velocity vector, namely

$$\boldsymbol{\imath} = -\boldsymbol{\nabla} U \tag{8.1}$$

The time independent regimes are determined by the condition

$$\boldsymbol{\nabla} U = \boldsymbol{0} \tag{8.2}$$

and the stability of these states are given by the criterion

$$\nabla^2 U \geq 0 \tag{8.3}$$

which is to be evaluated at a given non-equilibrium stationary state. The existence of U, like any other non-equilibrium thermodynamic potential, implies that the line integral of the velocity vector taken around a closed curve in α-space vanishes. This is entirely equivalent to the condition of exactness in the two-dimensional case. It implies that the non-equilibrium process is free of rotational motion.

When rotational motion exists, the non-equilibrium process cannot be described solely in terms of scalar non-equilibrium thermodynamic potentials. The existence of rotational motion can only be determined by the kinetic equations

$$\boldsymbol{\iota} = \boldsymbol{f}(\boldsymbol{\alpha}) \tag{8.4}$$

governing the motion. In order to show the connection between rotational motion and the destruction of the exactness conditions for a scalar potential, we integrate the time rate of change of the specific entropy between two definite times t_0 and t_1 (Lavenda, 1972*a*):

$$\Delta\dot{\eta} = \int_{t_0}^{t_1} \dot{\eta}\,\mathrm{d}t = -\int_{t_0}^{t_1} X_i \iota_i\,\mathrm{d}t = -\int_{\iota_i(t_0)}^{\iota_i(t_1)} X_i\,\mathrm{d}\iota_i \tag{8.5}$$

If $\dot{\eta}$ exists then it must be path independent, which implies that the variation of (8.5) vanishes. In this case, the change in $\Delta\dot{\eta}$ is a constant, dependent only on the initial and final states which are held fixed. The variation of (8.5) is

$$\delta\Delta\dot{\eta} = \int_{\iota_i(t_0)}^{\iota_i(t_1)} (\delta X_i\,\mathrm{d}\iota_i + X_i\,\delta\mathrm{d}\iota_i) \tag{8.6}$$

Integrating by parts and remembering that the limits of the integral are constants, we obtain

$$\delta\Delta\dot{\eta} = -\int_{\iota_i(t_0)}^{\iota_i(t_1)} (\partial_{\iota_j} X_i - \partial_{\iota_i} X_j)\delta\iota_j\,\mathrm{d}\iota_i - \int_{\iota_j(t_0)}^{\iota_j(t_1)} (\partial_{\iota_i} X_j - \partial_{\iota_j} X_i)\delta\iota_i\,\mathrm{d}\iota_j \tag{8.7}$$

where we have used the property that δ and d commute.

Whether or not the integrands in (8.7) vanish depends entirely on the kinetic equations (8.4), since the thermodynamic forces X_i are related to the internal state variables by the equations of state

$$X_i = -\partial_{\alpha_i} \eta \tag{8.8}$$

If

$$\partial_{\iota_j} X_i - \partial_{\iota_i} X_j = 0 \tag{8.9}$$

then

$$\delta\Delta\dot{\eta} = 0 \tag{8.10}$$

and $\dot{\eta}$ is a function of state that is necessarily independent of the path of the motion. Alternatively, if (8.9) does not vanish then the system is not free from rotation and new types of thermodynamic potentials (i.e. 'vector' potentials) are needed to describe the nonlinear thermodynamic process.

We may therefore classify the dynamic, cooperative phenomena according to whether the motion is irrotational or rotational. This distinction must only be made in nonlinear thermodynamics, since rotational motion is outlawed in the linear range. However, the type of irrotational or potential motion that is executed by system transitions among non-equilibrium stationary states can only occur in nonlinear thermodynamics.

We recall from chapter 2 that the Onsager reciprocal relations (8.9) are due to the detailed balancing that occurs at equilibrium. The conditions of detailed balancing, which equate every elementary process with its reverse, are more stringent than the stationarity conditions of the kinetic equations (8.4). The two sets of conditions must, however, be compatible so that the conditions of detailed balancing provide one time independent solution to the kinetic equations (8.4). If the kinetic equations are nonlinear, there may be other physically acceptable solutions to the kinetic equations. If we consider chemical reactions, then the conditions of detailed balancing at equilibrium yield the law of mass action. By a variation of the parameters in the law of mass action we can trace out what is referred to as a time independent 'branch' solution, which we shall refer to as the 'thermodynamic branch'. At equilibrium, the thermodynamic branch coincides with the conditions obtained from the detailed balancing of the elementary processes. Moreover, throughout the linear range, the thermodynamic branch is unique.

Other physically acceptable branch solutions to the time independent set of nonlinear kinetic equations can only appear in nonlinear thermodynamics (Lavenda, 1970). We shall refer to these branches as 'kinetic branches'. Once the kinetic branches have appeared, no single non-equilibrium stationary state can be globally stable. This is in direct contrast to linear thermodynamics, where the thermodynamic branch is unique. We can thus foresee the possibility of transitions between the thermodynamic and kinetic branches once the non-equilibrium stationary state on the thermodynamic branch becomes unstable.

In section 8.1, multistationary state transitions using the properties of the velocity potential are analysed. This analysis requires only a single dimension in α-space. When we enlarge the space to two dimensions (that is, the phase plane), there is no longer any guarantee that the velocity potential is single valued. Although we know that an integrating factor will always exist in the case of two state variables, we rather choose to associate the multivalued nature of the velocity potential with a particular form of the motion. This will provide only a kinematic description of the motion which we describe in section 8.2. In order to obtain information concerning the dynamics of rotational non-equilibrium processes, a thermodynamic analysis of the force fields is undertaken in the remaining sections of this chapter. It will become apparent that there is an underlying unity to all linear field theories irrespective of whether they occur in real space or phase space.

8.1 The velocity potential analysis of multistationary state transitions

Multistationary state transitions are the kinetic analogues of equilibrium second order phase transitions. A second order phase transition is one in which the thermodynamic potential and its first derivative are continuous while the second derivative is discontinuous. However, unlike equilibrium second order phase transitions, multistationary state transitions can occur without a change in the spatial symmetry of the system. And for simplicity of presentation we shall consider only those non-equilibrium transitions in which there is no change in the spatial symmetry of the system (Lavenda, 1970, 1972*b*). The theory of non-equilibrium transitions, involving a change in spatial symmetry, is somewhat more complex and it has been given by the author (Lavenda, 1975).

Time independent, non-equilibrium transitions require only a half-degree of freedom. The kinetic equations (8.4) can be reduced to a single equation of a half-degree of freedom by setting all but one of the internal state variables at their stationary state values. Through the process of elimination, we obtain the remaining velocity as a (possible) highly nonlinear function of its corresponding internal state variable, e.g.

$$\iota = F(\alpha) = \alpha^n + a_1\alpha^{n-1} + \ldots + a_{n-1}\alpha + a_n \tag{8.1.1}$$

where the a_i are time independent coefficients of the nth order polynomial. The velocity potential U is then constructed by integrating (8.1.1) with respect to α to give

$$U(\alpha) = -\int \iota(\alpha)\mathrm{d}\alpha + \text{const} \tag{8.1.2}$$

The physically admissible non-equilibrium stationary states are the real, positive roots of

$$\partial_\alpha U = 0: \quad F(\alpha) = 0 \tag{8.1.3}$$

The stability of a given non-equilibrium stationary state is determined by (8.3) which is the coefficient of the second order term in the series expansion of U about a given stationary state, namely

$$U = U_S + (\partial_\alpha U)_S\xi + \tfrac{1}{2}(\partial_\alpha^2 U)_s\xi^2 + \frac{1}{3!}(\partial_\alpha^3 U)_S\xi^3 + \ldots \tag{8.1.4}$$

where the subscript 's' denotes that the differential coefficients are to be evaluated at the stationary state and ξ is the deviation in α from its stationary state value.

At any given stationary state, the first order term in (8.1.4) vanishes and the criterion for stationary state stability is

$$\partial_\alpha^2 U > 0 \qquad \text{or} \qquad \partial_\alpha \iota < 0 \tag{8.1.5}$$

which has the form of a dynamic Le Châtelier criterion that we have discussed in section 6.1. It requires the velocity to be a decreasing function of its conjugate internal state variable in all stable processes. If U were an equilibrium thermodynamic potential, then the vanishing of the second order term in the series expansion (8.1.4) would indicate the critical point in a second order phase transition (Landau and Lifshitz, 1958). Stability is then governed by the lowest order, non-vanishing term in the series expansion (8.1.4). From symmetry considerations it can be shown that the odd order terms in the series expansion of the thermodynamic potential vanish and stability is governed by the fourth order term. In other words, the stability criteria of second order phase transitions are independent of the direction of the fluctuation. It will be seen that this is not the case for non-equilibrium transitions.

In non-equilibrium transitions, the vanishing of the second order term in (8.1.4) determines what is known in nonlinear mechanics as a 'bifurcation' point. Bifurcation points correspond to metastable states where stable and unstable stationary states coalesce. In other words, a point of bifurcation always separates two branch solutions of the time independent kinetic equations that have complementary stability characteristics. In the more general case, the solutions need not be time independent and we are then dealing with a multiplicity of periodic states (Lavenda, 1970).

In contrast to equilibrium second order phase transitions, there are no symmetry conditions and thus the stability of the motion is determined by the direction of the fluctuation. It is indeed logical that the evolution of the system should depend upon the direction of the fluctuations, since stable and unstable stationary regimes always arrange themselves in alternating fashion (Lavenda, 1972*b*). This constitutes a selection rule which can be used to predict the appearance and stability of stationary regimes. In analogy with electrostatics, we would associate stable and unstable stationary states with sinks and their corresponding sources. Therefore, at a bifurcation point, where stable and unstable states coalesce, there forms a 'hybrid' state.

If the fluctuation occurs in the domain governed by the stable stationary state, referred to as '*plage de stabilité*' (Ronsmans, 1964), then the system will eventually return to the metastable state. In the opposite case, where the fluctuation has occurred in the region controlled by the unstable stationary state, the system will evolve to another stationary state on a different branch solution. Hence, the velocity potential, unlike equilibrium thermodynamic potentials, will have a relative, instead of an absolute, minimum at any time independent state. In other words, when there exists more than one solution to the kinetic equations, no solution can be globally stable.

It appears that nature has introduced the kinetic branches in order to preserve system stability once the system arrives at a point of instability on the thermodynamic branch. In chemical kinetics this would mean that if there exists more than one physically acceptable solution to the kinetic equations, then the law of mass action cannot be continuously varied (that is, driving the system further from chemical equilibrium) without the system at some point becoming unstable. Instability induces the system to make a transition to a state on a stable kinetic branch. Moreover, instability cannot occur in linear thermodynamics, since the thermodynamic branch is unique. The question of the stability of a non-equilibrium process only arises in nonlinear thermodynamics.

8.2 Thermokinematics of rotational non-equilibrium processes

If we release the constraint that all but one of the internal state variables are maintained at their stationary state values, then there is no longer any guarantee that a velocity potential exists in nonlinear thermodynamics. As already mentioned, we can always find an integrating factor in the case of two internal state variables which will render dU an exact differential. However, it is not always possible to attribute a physical significance to the integrating factor. It therefore becomes of interest to ask whether any physical property of the non-equilibrium process can be attributed to the multivalued nature of non-equilibrium thermodynamic potentials.

Let us consider the case of two internal state variables in some detail. Suppose that a physically acceptable stationary state exists in the α_1, α_2-plane. The pair of kinetic equations (8.4) is then linearised about this stationary state and we obtain the variational equations

$$\dot{\xi}_1 = k_{11}\xi_1 + k_{12}\xi_2 \tag{8.2.1}$$

$$\dot{\xi}_2 = k_{21}\xi_1 + k_{22}\xi_2 \tag{8.2.2}$$

where the k_{ij} are the time independent rate constants. Since the velocities vanish at the stationary state, the total differential of the velocity potential is

$$\mathrm{d}U = -(\dot{\xi}_1 \mathrm{d}\xi_1 + \dot{\xi}_2 \mathrm{d}\xi_2) \tag{8.2.3}$$

Now there is no guarantee that (8.2.3) will be an exact differential, since the condition $k_{12} = k_{21}$ is not, in general, an Onsager relation. It will be an Onsager reciprocal relation only in the case that the stationary state is placed symmetrically in the phase plane. Hence, in general, we have to introduce a 'stretching' transformation (Lavenda, 1977)

$$\xi' = \mathbf{E}\xi \tag{8.2.4}$$

where the matrix of the transformation is

$$\mathbf{E} = \begin{pmatrix} |k_{21}|^{1/2} & 0 \\ 0 & |k_{12}|^{1/2} \end{pmatrix} \tag{8.2.5}$$

in order to render the cross coefficients symmetric or antisymmetric. This will depend upon whether k_{12} and k_{21} are of the same or opposite sign. That is, a strain of the form (8.2.4) will always bring out the latent symmetry in the variational equations (8.2.1) and (8.2.2). Thus, if k_{12} and k_{21} are of the same sign, we need not be concerned that the exactness conditions for U do not coincide with Onsager's principle.

The condition for the existence of a velocity potential can be formulated in a way which brings out the significance of the cross coefficient symmetry. By a translation, we can make the stationary state coincide with the origin of our phase plane. Then in the ξ_1, ξ_2-plane we construct a velocity vector with components $\dot{\xi}_1$ and $\dot{\xi}_2$ and integrate around a closed curve Γ. Using the planar polar coordinates $\xi_1 = r\cos\phi$ and $\xi_2 = r\sin\phi$ we obtain

$$\oint_\Gamma \dot{\xi}\cdot\mathrm{d}\xi = -\oint_\Gamma \mathrm{d}U = (k_{12} - k_{21})r^2\pi = \begin{cases} 0, & k_{12} = k_{21} \\ 2k_{12}r^2\pi, & k_{12} = -k_{21} \end{cases} \tag{8.2.6}$$

It will be appreciated that the condition for the vanishing of the line integral of the velocity vector, taken around a closed curve in the phase plane, is identical to the exactness condition which guarantees the existence of a velocity potential. When this integral does not vanish, we are dealing with a completely new form of the motion, for which the velocity potential is said to be multivalued. Each time the system completes a closed circuit, the velocity potential increases by a factor $2\pi r^2$, which in hydrodynamic terminology is known as the 'circulation'. The motion is no longer potential motion and the value of the line integral in (8.2.6) measures the intensity of rotation. In hydrodynamics, this form of motion is called a 'vortex' and we have seen that such motion is due to the action of non-conservative positional forces (cf. section 7.3).

Earlier, we associated the condition that the velocity potential must satisfy a Poisson equation of the form, cf. (8.3),

$$\nabla^2 U = -\boldsymbol{\nabla}\cdot\dot{\xi} = -(k_{11}+k_{22}) \geq 0 \tag{8.2.7}$$

with a criterion of stability. The inequality sign signifies that the system is devoid of all internal sources. In the special case where the equality sign applies, the velocity potential satisfies the Laplace equation. This indicates that there are no sinks or sources and such a state is metastable. It is then possible to define a conjugate function to the velocity potential, which is known in hydrodynamics as the 'stream' function. Denoting V as the stream function, its total differential is

$$\mathrm{d}V = (\dot{\xi}_2 \mathrm{d}\xi_1 - \dot{\xi}_1 \mathrm{d}\xi_2) \tag{8.2.8}$$

In terms of the stream function, the velocity components are

$$\partial_{\xi_1} V = \dot{\xi}_2 \quad \text{and} \quad \partial_{\xi_2} V = -\dot{\xi}_1 \tag{8.2.9}$$

The exactness condition for the stream function

$$\partial_{\xi_2}\dot{\xi}_2 = -\partial_{\xi_1}\dot{\xi}_1 \tag{8.2.10}$$

implies that the velocity potential satisfies the Laplace equation. Then, from equations (8.2.1), (8.2.2) and (8.2.8), we obtain a family of streamlines

$$V = \tfrac{1}{2}k_{21}(\xi_1^2 + \xi_2^2) = \text{const} \tag{8.2.11}$$

which are a family of concentric circles in the phase plane that are everywhere parallel to the flow. If we had not introduced the stretching transformation (8.2.4) then this family of closed curves would have been ellipses. If the stream function exists, then the streamlines cannot end abruptly in the interior of the system; rather, they must form closed curves such as (8.2.11) in the phase plane.

In nonlinear mechanics, the non-equilibrium stationary state is referred to as a 'centre'. Once the system has been perturbed from the stationary state, it will show neither the tendency to evolve to nor from the stationary state. Rather it will rotate about the stationary state indefinitely. This is analogous to Kelvin's theorem on the permanence of circulation. It states that the time derivative of the line integral of the velocity vector vanishes, namely

$$\mathrm{d}_t \oint_\Gamma \dot{\xi}\cdot\mathrm{d}\xi = 0 \tag{8.2.12}$$

or that the circulation is a constant of the motion. If we calculate the line integral of the velocity vector around a closed curve, we find

$$\oint_\Gamma \dot{\xi}\cdot\mathrm{d}\xi = |\boldsymbol{\omega}|2\pi r^2 = \text{const} \tag{8.2.13}$$

where $\boldsymbol{\omega}$ is the 'vorticity' vector

$$\boldsymbol{\omega} = \tfrac{1}{2}(\boldsymbol{\nabla}\wedge\dot{\xi}) \tag{8.2.14}$$

and the curl operator is defined in the phase plane. The vorticity vector serves as a measure of the circulation. We can cast (8.2.13) in a more suggestive form, using Stokes's theorem,

$$\oint_\Gamma \dot{\xi}\cdot d\xi = \int_S (\nabla \wedge \dot{\xi})\cdot d\boldsymbol{S} = \int_S 2\boldsymbol{\omega}\cdot d\boldsymbol{S} = |\boldsymbol{\omega}|2\pi r^2 \tag{8.2.15}$$

The surface integral has the physical meaning that it is the flow of the rotation vector through the surface S which is called the intensity of the vorticity. As we have already noted in (8.2.6), it is equal to the circulation.

Thus, if the non-equilibrium stationary state is a centre then the circulation is permanent. It expresses the fact that the rotational motion is constant for all times (that is, Kelvin's theorem). A nonlinear thermodynamic example is chemical rotations about a non-equilibrium stationary state (Bak, 1959). The non-equilibrium stationary state is metastable. However, in contrast to non-equilibrium transitions which use only a half-degree of freedom, the state of metastability now occurs at a finite frequency. Hence, we have two ways in which a non-equilibrium process can become metastable: (1) at zero frequency, indicative of non-equilibrium, transitions among stationary regimes, and (2) at finite frequency, which is characteristic of rotations about a non-equilibrium stationary state. Only in the first case do we have potential motion to which the classical analysis of nonlinear thermodynamics applies.

In general, a single-valued stream function will not exist. It is then of interest to ask about the significance of a multivalued stream function, as we did with the velocity potential. Evaluating the line integral of the stream function about a closed curve Γ in the phase plane, we obtain with the aid of the variational equations

$$\oint_\Gamma dV = -(k_{11}+k_{22})\pi r^2 \geq 0 \tag{8.2.16}$$

The multivalued nature of the stream function serves as a measure of the negative source strength in the non-equilibrium system. Inequality (8.2.16) is a criterion of stability that is entirely equivalent to the condition that the velocity potential satisfy the Poisson equation (8.2.7). The equality sign in (8.2.16) applies to the case where the velocity potential satisfies the Laplace equation, indicating that the non-equilibrium stationary state is metastable. Alternatively, if the stream function satisfies the Laplace equation

$$\nabla^2 V = 0 \qquad \text{or} \qquad 2\boldsymbol{\omega} = (\nabla \wedge \dot{\xi}) = \boldsymbol{0} \tag{8.2.17}$$

the motion is potential motion and there exists a single-valued velocity potential. In fact, condition (8.2.17) is none other than the exactness condition for the velocity potential.

In summary, we have performed an analysis of the velocity field in terms of the velocity potential and stream function. The result is the association of the elements of the variational equations with the multivalued nature of the velocity potential and stream function. This has allowed us not only to determine the behaviour of the motion but also the stability of the motion. A multivalued velocity potential indicates that the motion is rotational, while a multivalued stream function determines the stability of the motion.

Although analogies between non-equilibrium rotational processes and hydrodynamical vortex motion are fruitful, they cannot be pushed very far. The velocity field analysis is limited to the phase plane (that is, two dimensional flow) and it provides only information concerning the kinematics of the motion. In the next section we shall remedy these limitations through the development of a thermodynamic analysis of the force fields. It will then be appreciated that the thermodynamic analysis of the force fields is complementary to the velocity field analysis that we have developed in this section.

8.3 Thermodynamics of force fields

Field theories that treat matter as if it was continuously distributed throughout space instead of considering localised mass points are very common in macroscopic physics. The basic idea is that interactions among material components of the system occur through an intermediary known as the 'field'. The key to the construction of a thermodynamic field theory is to be found in Maxwell's field analysis of electrodynamics, that was built upon a model which treated the vacuum as an elastic solid. Although the field equations need no justification in terms of a particular model, we will find it convenient to conceive of nonlinear thermodynamic processes in terms of forces that act on an elastic solid (Lavenda, 1977).

First, the field theory is linear. This allows us to use Helmholtz's theorem to express the most complicated form of system motion as the sum of a translation, a deformation and a rotation of an imaginary elastic material that has been acted upon by a force which displaced it slightly from its equilibrium position. Since the non-equilibrium processes to be considered are spatially homogeneous, these motions will not occur in real space but rather in the phase space which is spanned by the internal state variables and their time-rates-of-change. The exclusion of spatially dependent processes is not a limitation of the theory, provided such processes are translationally invariant. For then, the spatial coordinates can be removed by a spatial Fourier transform. Hence, the phase space will consist of configuration and velocity spaces.

Our thermodynamic field theory conceives of the motion of nonlinear thermodynamic processes to be the result of system stresses that deform and rotate (that is, twist) an elastic material. Translatory motion is removed by shifting the origin of the phase space so that it coincides with the non-equilibrium stationary state under consideration.

Secondly, the basic equations fall into two classes: the thermodynamic field principles which describe the energetics of the field and the equations of motion which govern its dynamics. The thermodynamic field principles of the balance of power would correspond to the conservation principles of electrodynamics. The field equations, which are obtained through the generalised equations of state, express the system forces in terms of the generalised coordinates and velocities. In other words, the generalised equations of state convert the equations of motion into field equations. In addition, the motions in configuration and velocity spaces are interrelated through the equations of motion or the kinetic equations.

Thirdly, the phase space will, in general, contain $2N$ dimensions. Although the field operators of gradient, divergence and curl can be defined in such an abstract space, a more intuitive approach would be to limit the phase space to six dimensions in which the configuration and velocity spaces form separate Cartesian coordinate systems. We base such an approach on the premise that the behaviour of complex, nonlinear thermodynamic processes are the result of interactions among elementary or 'unit' processes. These elementary processes form the building blocks of the complex, nonlinear thermodynamic processes. In other words, we shall choose an interaction representation of nonlinear thermodynamic processes in the development of the thermodynamic field theory.

Finally, our approach will be inductive rather than the usual deductive nature of the classical field theories. This is to say, instead of determining the equations of motion from the known fields, we will determine the thermodynamic field equations from given sets of kinetic equations. Therefore, in order to write down a set of field equations and thermodynamic field principles, we must consider a classification of nonlinear thermodynamic forces. This can be accomplished by separating positional and motional forces into irrotational and solenoidal components. Here, we must be careful to distinguish between gyroscopic and circulatory forces, both of which cause rotational motion. In the previous section, the rotational motion was caused by circulatory forces and it is more correctly referred to as circulatory motion.

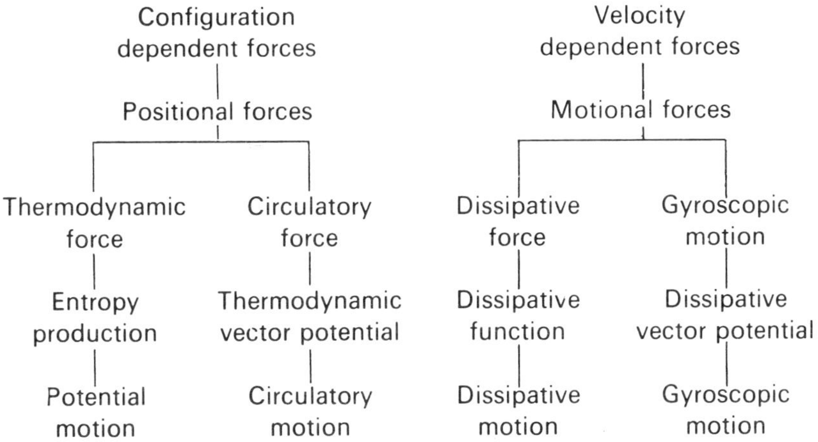

Figure 8.1 Classification of linear and non-linear thermodynamic forces

Figure 8.1 provides such a classification of nonlinear thermodynamic forces; there are two classes of system forces which are distinguishable on the basis of their functional dependencies. The columns on the left apply necessarily to linear thermodynamics, whereas the columns on the right apply only to nonlinear thermodynamics. Actually, if it were not for the solenoidal components of the positional and motional forces, a thermodynamic field theory would be entirely superfluous. Hence, there is really no need of a field theory in linear thermodynamics. In nonlinear thermodynamics, the concept of field is extremely useful, since it makes possible a description of the forces that do not act along straight lines of the system motion (Lavenda, 1977). These forces create rotational motion which can only occur in nonlinear thermodynamics.

8.4 Elements of the field theory

We consider the phase space to be comprised of configuration and velocity spaces, each of which forms a Cartesian coordinate system with unit vectors $\boldsymbol{i}, \boldsymbol{j}, \boldsymbol{k}$ that are pairwise orthogonal. This limitation on the phase space is not invoked out of necessity but because (*a*) we have chosen an interaction representation of complex, nonlinear thermodynamic processes, and (*b*) it will allow a direct comparison with the field theories of macroscopic physics which describe how the field quantities change in space and time.

The generalised coordinate and velocity vectors are defined as

$$\xi = \xi_1 \boldsymbol{i} + \xi_2 \boldsymbol{j} + \xi_3 \boldsymbol{k} \tag{8.4.1}$$

and

$$\dot{\xi} = \dot{\xi}_1 \boldsymbol{i} + \dot{\xi}_2 \boldsymbol{j} + \dot{\xi}_3 \boldsymbol{k} \tag{8.4.2}$$

respectively. The linearised kinetic or variational equations are given in the form

$$\mathbf{R}\dot{\xi} = \mathbf{S}\xi \tag{8.4.3}$$

for systems comprised of half-degrees of freedom. Our thermodynamic field analysis will permit us to treat a system comprised of three half-degrees of freedom or, at most, a system described by six variational equations which have been converted into three variational equations of single degrees of freedom. We shall take up each of these representations of the variational equations in turn.

The fundamental operators in the configuration and velocity spaces are the nablas

$$\boldsymbol{\nabla} = \boldsymbol{i}\partial_{\xi_1} + \boldsymbol{j}\partial_{\xi_2} + \boldsymbol{k}\partial_{\xi_3} \tag{8.4.4}$$

and

$$\dot{\boldsymbol{\nabla}} = \boldsymbol{i}\partial_{\dot{\xi}_1} + \boldsymbol{j}\partial_{\dot{\xi}_2} + \boldsymbol{k}\partial_{\dot{\xi}_3} \tag{8.4.5}$$

respectively. The thermodynamic and dissipative forces can be defined using these operators, namely

$$\chi_T = -\boldsymbol{\nabla}\eta \tag{8.4.6}$$

$$\chi_D = -\dot{\boldsymbol{\nabla}}\phi \tag{8.4.7}$$

The exactness conditions for these two non-equilibrium thermodynamic potentials are expressible in the form

$$\boldsymbol{\nabla} \wedge \chi_T = 0 \tag{8.4.8}$$

$$\dot{\boldsymbol{\nabla}} \wedge \chi_D = 0 \tag{8.4.9}$$

Expressed in words: if the forces are gradients of scalar potentials then the curl of the gradient of any scalar function is zero. When the exactness conditions (8.4.8) and (8.4.9) are invalidated by the motion, that is described by the variational equations (8.4.3), the forces can no longer be derived solely from scalar potentials. Therefore, it is the forces and not the non-equilibrium potentials which are of primary importance in the thermodynamic field analysis.

According to a fundamental theorem, first proved by Stokes, any continuous vector field χ_i, defined everywhere in space and vanishing at infinity together with its first derivative, can be expressed as the sum of an irrotational field χ_i' and a solenoidal field χ_i'', namely

$$\chi_i = \chi_i' + \chi_i'' \tag{8.4.10}$$

where

$$\chi_T' = -\nabla \tfrac{1}{2}\delta^2\eta; \quad \chi_D' = -\dot{\nabla}\phi \tag{8.4.11}$$

$$\chi_T'' = -(\nabla \wedge A); \quad \chi_D'' = -(\dot{\nabla} \wedge B) \tag{8.4.12}$$

A and B represent vector potentials. The thermodynamic and dissipative force variations can therefore be expressed as

$$\chi_T = \tfrac{1}{2}\nabla(\xi \cdot \chi_T) - (\nabla \wedge A) \tag{8.4.13}$$

and

$$\chi_D = \tfrac{1}{2}\dot{\nabla}(\dot{\xi} \cdot \chi_D) - (\dot{\nabla} \wedge B) \tag{8.4.14}$$

respectively. A is called the 'thermodynamic' vector potential and it is defined as

$$A = \tfrac{1}{2}(\xi \wedge \chi_T'') \tag{8.4.15}$$

while the definition of the 'dissipative' vector potential B is

$$B = \tfrac{1}{2}(\dot{\xi} \wedge \chi_D'') \tag{8.4.16}$$

The terminology is not indicative of the character of the vector potentials; rather, it serves only as a reminder that the vector potentials are defined in terms of the solenoidal components of the forces whose irrotational components are the thermodynamic and dissipative forces.

Let us now return to our imaginary elastic material and consider the forces that cause deformation and rotation. The irrotational components of the force fields cause deformation, while rotation is due to the action of the solenoidal components. What we mean by rotation is the twisting of the elastic solid and not the rotation of the body as a whole. The stresses which act in configuration space can be expressed in dyadic notation as

$$\nabla\chi_T = \tfrac{1}{2}(\nabla\chi_T + \chi_T\nabla) - \tfrac{1}{2}(\nabla \wedge \chi_T) \wedge \mathbf{I} \tag{8.4.17}$$

while the stresses that act in velocity space are

$$\dot{\nabla}\chi_D = \tfrac{1}{2}(\dot{\nabla}\chi_D + \chi_D\dot{\nabla}) - \tfrac{1}{2}(\dot{\nabla} \wedge \chi_D) \wedge \mathbf{I} \tag{8.4.18}$$

The dyadic $\nabla\chi_T$ at any point in configuration space is determined by nine components along the orthogonal axes. The symmetric part of the dyadic is one-half the sum of the dyadic

$$\nabla\chi_T = \boldsymbol{i}\partial_{\xi_1}\chi_T + \boldsymbol{j}\partial_{\xi_2}\chi_T + \boldsymbol{k}\partial_{\xi_3}\chi_T \tag{8.4.19}$$

and its conjugate

$$\chi_T\nabla = \partial_{\xi_1}\chi_T\boldsymbol{i} + \partial_{\xi_2}\chi_T\boldsymbol{j} + \partial_{\xi_3}\chi_T\boldsymbol{k} \tag{8.4.20}$$

Similar forms are obtained for the symmetric part of the stress dyadic in velocity space.

Whereas the stress dyadic (8.4.17) operates on the generalised coordinate vector to produce the positional force χ_T, the stress dyadic (8.4.18) operates on the velocity vector to give the motional force χ_D. These forces are the generalisations of the thermodynamic and dissipative force variations in nonlinear thermodynamics. The symmetric part of the stress dyadic causes deformations in the elastic solid that are in the normal and tangential directions to its surface. A consideration of the torques on an element of volume shows that the symmetric part of the stress dyadic does not cause rotation. In linear thermodynamics there can be no rotation and the symmetry manifested by the stress dyadic in velocity space is expressed by Onsager's reciprocal relations. However, in nonlinear thermodynamics there can be antisymmetric components of the stress dyadics $\nabla\chi_T$ and $\nabla\chi_D$. These components cause circulatory and gyroscopic motions, respectively. In a mechanical sense, non-equilibrium processes that occur in linear thermodynamics are said to be in 'static' equilibrium, since circulatory and gyroscopic motions are absent.

The positional and motional forces are derivable from their stress dyadics in the following way:

$$\chi_T = \xi\cdot\nabla\chi_T = \tfrac{1}{2}\xi\cdot(\nabla\chi_T+\chi_T\nabla)-\tfrac{1}{2}\xi\wedge(\nabla\wedge\chi_T) \tag{8.4.21}$$

$$\chi_D = \dot{\xi}\cdot\dot{\nabla}\chi_D = \tfrac{1}{2}\dot{\xi}\cdot(\dot{\nabla}\chi_D+\chi_D\dot{\nabla})-\tfrac{1}{2}\dot{\xi}\wedge(\dot{\nabla}\wedge\chi_D) \tag{8.4.22}$$

The first members on the right are the thermodynamic and dissipative force variations. The second members are the circulatory and gyroscopic forces. As we have seen in section 7.4, the forces that cause rotational motions are distinguishable according to their effects on the stability of the motion. That is, although the circulatory forces χ''_T do no real work, they nevertheless have a first order effect upon stability. The character of the circulatory force likens to an electromotive force in electrodynamics. On the contrary, gyroscopic forces were shown not to have a first order effect upon the stability of the motion. Their presence depends upon the choice of the reference frame. Examples of such forces are the Coriolis force in a rotating reference frame and the magnetic forces due to electric currents.

Taking the inner product of the positional force (8.4.21) with the generalised coordinate vector, we obtain

$$\tfrac{1}{2}\delta^2\eta = -\tfrac{1}{2}\xi\cdot(\nabla\chi_T+\chi_T\nabla)\cdot\xi \tag{8.4.23}$$

which shows that the circulatory forces do no real work on the system. The inner product of the motional force with the velocity vector gives

$$\phi = \tfrac{1}{2}\dot{\xi}\cdot(\dot{\nabla}\chi_D+\chi_D\dot{\nabla})\cdot\dot{\xi} \tag{8.4.24}$$

expressing the fact that there is no rate of working due to gyroscopic forces. The next step is to relate the positional and motional force fields to the variational equations. The resulting field equations will permit us to derive the thermodynamic field principles that describe the internal laws of the field.

8.5 Thermodynamic principles of the field

The variational equations of the internal state variables (8.4.3) can be written as the variational force balance law

$$\chi_D - \chi_T = 0 \tag{8.5.1}$$

by employing the generalised equations of state

$$\chi_D = -\mathbf{R}\dot{\xi} \tag{8.5.2}$$

$$\chi_T = \mathbf{S}\xi \tag{8.5.3}$$

Whether or not the forces can be derived from the scalar, non-equilibrium thermodynamic potentials depends on the symmetries of the coefficient matrices, **R** and **S**.

If both coefficient matrices are symmetric, as they must be in linear thermodynamics, then the inner product of the force variational equation (8.5.1) with the velocity vector gives the linear thermodynamic principle

$$2\phi - \tfrac{1}{2}\delta^2\dot{\eta} = 0 \tag{8.5.4}$$

Now, in nonlinear thermodynamics, there is no guarantee that the phenomenological coefficient matrices are symmetric and we can consider the limiting case in which there is permanent circulation about a non-equilibrium stationary state. In this case

$$\mathbf{S} = -\mathbf{S}^T \tag{8.5.5}$$

and the irrotational component of the positional force field vanishes. There remains only the circulatory force

$$\chi''_T = -(\nabla \wedge A) \tag{8.5.6}$$

The inner product of the force variational equation (8.5.1) with the velocity vector now gives

$$2\phi = \dot{\xi}\cdot(\nabla \wedge A) \tag{8.5.7}$$

Using the vector equation

$$\dot{\xi}\cdot(\nabla \wedge A) = -\nabla\cdot(\dot{\xi} \wedge A) \tag{8.5.8}$$

the thermodynamic field principle of the balance of power (8.5.7) can be written as

$$2\phi = -\nabla\cdot \boldsymbol{P} \tag{8.5.9}$$

where

$$\boldsymbol{P} = (\dot{\xi} \wedge A) \tag{8.5.10}$$

The vector $\boldsymbol{P}$ is the energy flux density in configuration space (Lavenda, 1977); it is the mean field energy per unit time. The energy flux vector bears a formal resemblance to Poynting's vector in electrodynamics. The positive semidefinite form of the dissipation function demands that the energy flux be directed inward. In other words, the rate at which energy is being dissipated is compensated exactly by the energy influx. This is how a state of permanent circulation about a non-equilibrium stationary state is maintained energetically.

In the most general case, we have

$$\mathbf{S} = \tfrac{1}{2}(\mathbf{S}+\mathbf{S}^T) + \tfrac{1}{2}(\mathbf{S}-\mathbf{S}^T) \tag{8.5.11}$$

in place of (8.5.5), since there can be no gyroscopic motion in non-equilibrium processes that are described by variational equations of half-degrees of freedom. The coefficient matrix **R** will therefore always be symmetric and the thermodynamic field principle is

$$2\phi - \tfrac{1}{2}\delta^2\dot{\eta} = -\nabla\cdot \boldsymbol{P} \tag{8.5.12}$$

in its most general form. Equation (8.5.12) states that the energy influx is equal to the difference between the dissipation function and the excess entropy production. The maximum energy influx occurs in the case of permanent circulation for which the excess entropy production vanishes, that is (8.5.9). In general, the production of entropy in the system will decrease the energy influx. This brings out the energetical distinction between the dissipation function and the entropy production.

Although the thermodynamic field principles of non-equilibrium processes that are described by single degree of freedom variational equations are formally identical with those of half-degrees of freedom, the energetics are completely different. For non-equilibrium processes that are described in terms of single degrees of freedom, the generalised equation of state (8.5.3) is replaced by

$$\underset{T}{\chi} = \mathbf{S}\xi + \mathbf{M}\ddot{\xi} \tag{8.5.13}$$

as we have seen in section 5.2. In addition to circulatory forces, which are associated with the antisymmetric part of the coefficient matrix **S**, there can also be gyroscopic motion, which is indicated by the antisymmetric part of the generalised resistance matrix **R**. That is, **R** will in general not be symmetric and we have

$$\mathbf{R} = \tfrac{1}{2}(\mathbf{R}+\mathbf{R}^T) + \tfrac{1}{2}(\mathbf{R}-\mathbf{R}^T) \tag{8.5.14}$$

However, the presence of gyroscopic motion will never be indicated in a thermodynamic field principle, since the gyroscopic forces cause no rate of working. This is significant of the fact that gyroscopic motion does not have a first order effect upon the stability of the motion (cf. section 7.4). Hence, it does not enter into a thermodynamic field principle that describes the energetics of the nonlinear thermodynamic process. Gyroscopic motion can, however, be accounted for in the description of fields by variational principles as we shall show in the next section.

Prior to doing so, it is interesting to show the contrast between the energetics of systems that are composed of half-degrees and single degrees of freedom. Although both types of non-equilibrium processes can be described by the thermodynamic field principle (8.5.12), only in the single degree of freedom case can we set $\mathbf{R} = \mathbf{0}$, thereby obtaining

$$\tfrac{1}{2}\delta^2\dot{\eta} = \nabla\cdot \boldsymbol{P} \tag{8.5.15}$$

This thermodynamic field principle can be contrasted with (8.5.9) which applies to the limiting case of permanent circulation in nonlinear thermodynamic processes that are described by half-degrees of freedom. From the fact that the excess entropy production is positive in all stable systems, we conclude that there is an energy efflux. That is to say, the field not only possesses a certain energy density but it also gives rise to a flow of energy. Dissipation can only decrease the flow of energy. The difference between non-equilibrium processes that use half-degrees and single degrees of freedom is that the 'inertial' component of the thermodynamic force produces a positive rate of working. This is reflected in the thermodynamic field principle (8.5.15) in which there is an energy efflux.

8.6 Description of fields by thermodynamic variational principles

At this point it is worth while to summarise our field analysis. We began with a given set of variational equations. These equations govern the motion about non-equilibrium stationary states. The field equations were then obtained by introducing the force fields into the variational equations through the generalised equations of state. Finally, the thermodynamic field principles were obtained by forming the inner product of the force variational equation (8.5.1) with the velocity vector. The thermodynamic field principles are analogous to the conservation principles of electrodynamics.

We would now like to be able to derive the field equations from a thermodynamic variational principle. This will be possible in all cases analysed so far, save that of gyroscopic motion. Gyroscopic motion requires a mechanical Lagrangian formulation which will be discussed in due course.

It will be recalled that the free thermodynamic variational principle of linear thermodynamics is (cf. section 6.2)

$$\mathscr{L} = \phi(\dot{\xi}, \dot{\xi}) - \dot{\eta}(\dot{\xi}) = \min \tag{8.6.1}$$

Moreover, it will also be recalled that the variational principle uses the thermodynamic convention of performing all kinematically permissible variations in the velocities for a prescribed configuration. The stationarity condition for the thermodynamic Lagrangian gives the variational equations in the form of the force balance law (8.5.1). The thermodynamic variational principle is valid throughout linear thermodynamics; it may not be valid for nonlinear thermodynamic processes.

Since rotational motion can occur in nonlinear thermodynamics, we cannot expect that the thermodynamic Lagrangian will always be formed from scalar non-equilibrium thermodynamic potentials. In order to construct the appropriate nonlinear thermodynamic Lagrangian, we must return to the principle of least dissipation of energy (6.2.7) and use the thermodynamic field principle as the constraint. The constraint imposed by the thermodynamic field principle (8.5.12) can be explicitly introduced into the principle of least dissipation of energy through the method of the Lagrange undetermined multiplier. We then obtain

$$\mathscr{L} = \phi - \lambda(2\phi - \tfrac{1}{2}\delta^2\dot{\eta} + \boldsymbol{\nabla}\cdot\boldsymbol{P}) = \min \tag{8.6.2}$$

where λ is the undetermined multiplier. The variational expression (8.6.2) is equivalent to (6.2.7), since all we have done is to add zero. The stationary value of (8.6.2) with respect to variations in the velocity vector is

$$(1-\lambda)\partial_{\dot{\xi}}\phi - \lambda(\partial_{\dot{\xi}}\phi - \partial_{\dot{\xi}}\tfrac{1}{2}\delta^2\dot{\eta} - \nabla \wedge \boldsymbol{A}) = \boldsymbol{0} \tag{8.6.3}$$

The second member vanishes identically since it is just the field equation

$$\chi'_D - \chi'_T + \nabla \wedge \boldsymbol{A} = 0 \tag{8.6.4}$$

and hence $\lambda = 1$. The free variational principle of least dissipation of energy is

$$\mathscr{L} = \phi - \tfrac{1}{2}\delta^2\dot{\eta} + \nabla\cdot\boldsymbol{P} = \min \tag{8.6.5}$$

It will be observed that the stationary value of (8.6.5) yields the field equation (8.6.4). Its inner product with the velocity vector gives us back the thermodynamic field law (8.5.12). The extremum property of the thermodynamic Lagrangian follows, as usual, from the positive semidefinite form of the dissipation function.

As was previously remarked, the thermodynamic Lagrangian formulation cannot be applied to gyroscopic motion, since it cannot be accounted for in an equation that describes the rates of change of the energy densities and flows. The field equation for gyroscopic motion is

$$\chi_T = -\dot{\nabla} \wedge \boldsymbol{B} \tag{8.6.6}$$

It is easily seen that the inner product of the force variational equation (8.6.6) with the velocity vector vanishes. Therefore, in order to include gyroscopic motion within the limits of the field theory, we must return to a mechanical Lagrangian formulation.

The general method for converting a thermodynamic variational principle into a mechanical variational principle is to take the Legendre transform of the second variation of the specific entropy with respect to the velocities. Since gyroscopic motion occurs in systems comprised of single degrees of freedom, the specific entropy will be the sum of Gibbsian and non-Gibbsian parts. The velocities now belong to the set of state variables, cf. equation (5.2.22). It will be recalled from section 5.2 that the second variation of the specific entropy is a quadratic form in the velocities and displacements. Consequently, its total Legendre transform with respect to the velocities does not vanish, namely

$$\Lambda = \partial_{\dot{\xi}}\tfrac{1}{2}\delta^2\eta\cdot\dot{\xi} - \tfrac{1}{2}\delta^2\eta \tag{8.6.7}$$

Λ has the form of a mechanical Lagrangian which must be supplemented by a field term in order to account for gyroscopic motion. Defining $\boldsymbol{N}$ as the energy flux in velocity space

$$\boldsymbol{N} = \tfrac{1}{2}(\xi \wedge \boldsymbol{B}) \tag{8.6.8}$$

the mechanical Lagrangian, that leads to the correct field equation for gyroscopic motion, is

$$L = \Lambda + \dot{\nabla}\cdot\boldsymbol{N} = \text{extremum} \tag{8.6.9}$$

The variational principle (8.6.9) is to be interpreted in the mechanical sense: all kinematically permissible variations in the velocities and displacements are to be performed. Let us emphasise the fact that there is absolutely no connection between the mechanical Lagrangian (8.6.9) and the principle of least dissipation of energy. The mechanical Lagrangian (8.6.9) is only required to be an extremum along the actual path of the motion

$$(\mathbf{M}+\mathbf{M}^T)\ddot{\xi}+(\mathbf{R}-\mathbf{R}^T)\dot{\xi}+(\mathbf{S}+\mathbf{S}^T)\xi = 0 \tag{8.6.10}$$

which follows from the condition of stationarity of the mechanical Lagrangian (8.6.9). A thermodynamic principle of the balance of power cannot be obtained from the variational equations (8.6.10) by multiplying by the velocity vector, since the trace of the product of a symmetric with an antisymmetric matrix vanishes.

As a final comment, we would like to emphasise that the field equations and the thermodynamic field principles apply only to small motions about non-equilibrium stationary states. Although the field analyses have been performed in phase space, rather than the real three-dimensional space, the structure of the field equations bear a striking resemblance with those of hydrodynamics and electrodynamics. This is not surprising since all linear field theories can be reduced to the analyses of stresses that act on an elastic material. We have already noted that this is how Maxwell constructed his field theory of electrodynamics. Maxwell's equations are linear, as are the variational equations that have been used throughout this chapter. Need the field equations always be linear? The variational equations apply only to small motions about a non-equilibrium stationary state. They are linearised versions of more general equations which cover a much wider domain of the system motion. This would lead us to believe that linear field equations are only linearised versions of more general nonlinear laws.

References

Bak, T. A. (1959). On the existence of oscillating chemical reactions. *Bull. Acad. Roy. Belg. Cl. Sci.,* **45,** 116–129.

Landau, L. D. and Lifshitz, E. M. (1958). *Statistical Physics,* Pergamon Press, Oxford.

Lavenda, B. H. (1970). Ph. D. Thesis, University of Brussels.

Lavenda, B. H. (1972*a*). Generalized thermodynamic potentials and universal criteria of evolution. *Lett. Nuovo Cimento,* **3,** 385–390.

Lavenda, B. H. (1972*b*). Theory of multistationary state transitions and biosynthetic control processes. *Q. Rev. Biophys.,* **5,** 429–479.

Lavenda, B. H. (1975). Dynamic phase transitions in nonequilibrium processes. *Phys. Rev. A,* **11,** 2066–2078.

Lavenda, B. H. (1977). A field analysis of nonlinear irreversible thermodynamic processes. *Found. Phys.* **7,** 11/12, 906–926.

Ronsmans, P. (1964). Sur la stabilité et index des singularités multiples d'un système différential autonome. *Bull. Acad. Roy. Belg. Cl. Sci.,* **50,** 142.

9 Continuum Thermodynamics

A unification of phenomena from diverse branches of macroscopic physics is undertaken in this chapter. We span the time period between the classic inquiries of Eckart (1940) into the irreversible thermodynamics of fluid mixtures, to the recent and elegant formulation of the averaged Lagrangian approach to nonlinear dispersive waves by Whitham (1974). This is made possible through the construction of an internal state variable representation of the thermodynamics of the continua.

The crucial point in the thermodynamic analysis is the interpretation of internal dissipation in terms of the internal state or 'hidden' variables. Our approach can be contrasted with Truesdell's purely mechanical considerations of diffusion phenomena (Truesdell, 1957), which are based on the intuitive ideas of kinetic theory. Other works on the theory of mixtures, that follow similar lines of reasoning, can be found in Kelly (1964), Green and Naghdi (1965, 1967), Bowen (1967) and Müller (1968). Important distinctions can also be made with the methodology of classical non-equilibrium thermodynamics that constructs an expression for the entropy production (de Groot and Mazur, 1962) which is both logically and structurally incoherent.

In this chapter we arrive at the present-day frontiers of non-equilibrium thermodynamics. Having traced the development from a linear to a nonlinear thermodynamic theory, we now realise the need to construct a self-consistent theory of the thermodynamics of nonlinear irreversible processes. Section 7.6 suggested how such a theory could be constructed using averaging techniques borrowed from nonlinear mechanics. Here, we study nonlinear irreversible processes of the continua and illustrate how the interaction between nonlinear dissipation and linear dispersion can lead to the creation of a totally new phenomenon that could not have been predicted from a linear analysis, whether kinetic or thermodynamic. This provides the motivation for the development of a truly nonlinear thermodynamic theory of the continua. This, it is felt, will come about through a synthesis of thermodynamic and kinetic concepts which, in fact, has been the theme of this book.

9.1 The balance equations of the continua

The balance equations of the continua can be found in most, if not all, the standard texts on non-equilibrium thermodynamics (for example de Groot and Mazur, 1962; Fitts, 1962). We shall not write down the most general form of the balance equations but rather those equations that will be of use in later sections. In most respects, the notation of Truesdell (1969) is followed.

In the thermodynamics of mixtures, it is necessary to distinguish between the barycentric velocity $\dot{\boldsymbol{x}}$ and the so-called peculiar velocities $\grave{\boldsymbol{x}}_i$, where

$$\grave{\boldsymbol{x}}_i = \partial_t \boldsymbol{x}_i \tag{9.1.1}$$

The total mass density is the sum of all the partial mass densities, namely

$$\rho = \sum_i \rho_i \tag{9.1.2}$$

The definition of the barycentric velocity in terms of the peculiar velocities is

$$\dot{\boldsymbol{x}} = \frac{1}{\rho} \sum_i \rho_i \grave{\boldsymbol{x}}_i \tag{9.1.3}$$

The velocity relative to the mean motion or the so-called 'diffusion' velocity is defined as

$$\boldsymbol{u}_i = \grave{\boldsymbol{x}}_i - \dot{\boldsymbol{x}} \tag{9.1.4}$$

Although we require the total mass to be conserved in the system

$$\partial_t \rho + \nabla \cdot \rho \dot{\boldsymbol{x}} = \dot{\rho} + \rho \nabla \cdot \dot{\boldsymbol{x}} = 0 \tag{9.1.5}$$

the partial specific densities may not be conserved, namely

$$\grave{\rho}_i + \rho_i \nabla \cdot \grave{\boldsymbol{x}}_i = \partial_t \rho_i + \nabla \cdot \rho_i \grave{\boldsymbol{x}}_i = \rho \sigma_i \tag{9.1.6}$$

where σ_i is the source term that may be due to physical transfers or chemical reactions. On account of (9.1.5) we have

$$\sum_i \sigma_i = 0 \tag{9.1.7}$$

Similar arguments hold true for the partial specific energies and momenta.

We have already written down the equation of motion of the centre of mass

$$\rho \ddot{\chi} = \nabla \cdot \mathbf{T} + \rho \boldsymbol{b} \tag{3.1.4}$$

where we recall that $\mathbf{T}$ is the stress tensor and $\boldsymbol{b}$ are the body forces or, more generally, the external forces. The balance equation for the internal energy (3.3.36) is modified to include the rate of change of the kinetic energy (Eckart, 1940)

$$\mathrm{d}_t \int_V \tfrac{1}{2} \rho \dot{x}^2 \, \mathrm{d}V = \int_V \{\dot{\boldsymbol{x}} \cdot (\nabla \cdot \mathbf{T}) + \rho \dot{\boldsymbol{x}} \cdot \boldsymbol{b}\} \, \mathrm{d}V \tag{9.1.8}$$

Energy changes due to diffusion and local stresses $\mathbf{t}^{(i)}$, exerted by the individual constituents, are taken into account in the energy flux vector

$$\boldsymbol{J}_E = \boldsymbol{q} + \rho_i \varepsilon_i \boldsymbol{u}_i - \mathbf{t}^{(i)} \boldsymbol{u}_i \tag{9.1.9}$$

The stresses exerted by the individual constituents in the mixture have been designated as the 'peculiar' stresses by Truesdell (1969) in an apparent analogy with the peculiar velocities (9.1.1). Although the peculiar stresses $\mathbf{t}^{(i)}$ need not be symmetric, their sum

$$\mathbf{T} = \sum_i \mathbf{t}^{(i)} \tag{9.1.10}$$

is required to be symmetric by the Cauchy second law of motion, $\mathbf{T} = \mathbf{T}^T$ (Eringen and Ingram, 1965). If we consider an ideal fluid mixture then $\mathbf{t}^{(i)} = -p_i \mathbf{I}$, where p_i is the partial pressure of the constituent, and (9.1.10) reduces to Dalton's law of partial pressures. Moreover, if the need arose, we could also express the heat flux vector $\boldsymbol{q}$ and the body force $\boldsymbol{b}$ as the sum of constituent contributions. For our purposes, this will not be necessary and the interested reader is advised to consult Eringen and Ingram (1965), Ingram and Eringen (1967) and Truesdell (1969).

The balance equation for the energy is

$$\begin{aligned} &\mathrm{d}_t \int_V (\tfrac{1}{2}\rho \dot{x}^2 + \rho\varepsilon)\mathrm{d}V + \int_{\partial V} \{\boldsymbol{q} + \rho_i \varepsilon_i \boldsymbol{u}_i - \mathbf{t}^{(i)} \boldsymbol{u}_i\} \cdot \mathrm{d}\mathbf{s} \\ &= \int_{\partial V} \mathbf{T}\cdot\dot{\boldsymbol{x}}\cdot\mathrm{d}\boldsymbol{s} + \int_V \rho\theta\pi \mathrm{d}V \end{aligned} \tag{9.1.11}$$

In order to obtain a balance equation for the specific internal energy, we subtract (9.1.8) from (9.1.11). Then, since the volume V is arbitrary, we obtain

$$\rho\dot{\varepsilon} + \nabla\cdot(\boldsymbol{q} + \rho_i \varepsilon_i \boldsymbol{u}_i - \mathbf{t}^{(i)} \boldsymbol{u}_i) = \mathbf{T} : \nabla \dot{\boldsymbol{x}} + \rho\theta\pi_T \tag{9.1.12}$$

where π_T is the total power

$$\pi_T = \pi + \frac{1}{\theta}\dot{\boldsymbol{x}}\cdot\boldsymbol{b} \tag{9.1.13}$$

In the derivation of the local balance equation (9.1.12), we have used the property that if F is any function of space and time then

$$\mathrm{d}_t \int_V \rho F \mathrm{d}V = \int_V \rho \dot{F} \mathrm{d}V \tag{9.1.14}$$

which follows from the application of the divergence theorem and the conservation of total mass (9.1.5).

The balance equation for the specific entropy will also be modified due to a change in entropy that is caused by diffusion processes. The entropy flux vector will now contain two contributions

$$\boldsymbol{J}_S = \boldsymbol{q}/\theta + \rho_i \eta_i u_i \tag{9.1.15}$$

one due to the flow of heat and the other to diffusion. In place of the entropy balance equation (5.1.1), we now have

$$\rho\dot{\eta} + \nabla\cdot(\boldsymbol{q}/\theta + \rho_i \eta_i \boldsymbol{u}_i) = \rho\sigma_S \tag{9.1.16}$$

With the aid of the balance equations for the specific internal energy and entropy, we can now proceed to derive the generalised power equation.

9.2 The generalised power equation

The generalised power equation is derived by eliminating the divergence of the heat flux vector between equations (9.1.12) and (9.1.16). We then obtain

$$\rho(\dot{\varepsilon}-\theta\dot{\eta})+\nabla\cdot\{\rho_i u_i(\varepsilon_i-\theta\eta_i)-\mathbf{t}^{(i)}\cdot\boldsymbol{u}_i\} = -\boldsymbol{J}_s\cdot\nabla\theta+\mathbf{T}:\nabla\dot{\boldsymbol{x}}+\rho\theta(\pi_T-\sigma_S) \tag{9.2.1}$$

It will prove convenient to introduce the specific Helmholtz free energy ψ and the specific partial free energies ψ_i into the generalised power equation (9.2.1). The result is

$$\rho(\dot{\psi}+\eta\dot{\theta})+\nabla\cdot\{(\rho_i\psi_i\mathbf{I}-\mathbf{t}^{(i)})\boldsymbol{u}_i\} = -\boldsymbol{J}_S\cdot\nabla\theta+\mathbf{T}:\nabla\dot{\boldsymbol{x}}+\rho\theta(\pi_T-\sigma_S) \tag{9.2.2}$$

If we put $\pi = 0$ and interpret (9.2.2) as a Clausius – Duhem inequality then we find agreement with Eringen and Ingram (1965) but not with Truesdell (1969). In Truesdell's expression, the rate of working of the body forces against diffusion is lacking. The reason for the disagreement is that Truesdell did not take into account the balance equation for the kinetic energy due to the centre mass motion (9.1.8). Truesdell (1960) did note, however that the tensor $(\rho_i\psi_i\mathbf{I}-\mathbf{t}^{(i)})$ constitutes a generalisation of the ordinary definition of the chemical potential μ_i. The chemical potential tensor $\mathbf{C}^{(i)}$ is defined as (Truesdell, 1969)

$$\mathbf{C}^{(i)} = \psi_i\mathbf{I}-\frac{1}{\rho_i}\mathbf{t}^{(i)} \tag{9.2.3}$$

and it is easily shown to reduce to the ordinary definition of the chemical potential in the case of an ideal mixture. For an ideal mixture, $\mathbf{t}^{(i)} = -p_i\mathbf{I}$ and (9.2.3) becomes

$$\mu_i = \psi_i+p_i/\rho_i \tag{9.2.4}$$

With the power still equal to zero we can interpret (9.2.2) from the standpoint of classical non-equilibrium thermodynamics (cf. de Groot and Mazur, 1962). The assumption of local equilibrium is made and the equilibrium Gibbs relation is introduced into (9.2.2) through the expression for the substantial derivative of the specific free energy. Then a sequence of algebraic manipulations is used to express the entropy production as a sum of bilinear terms involving forces and fluxes. Specifically, one tries to eliminate the divergence term in (9.2.2), since it obviously has no business in the expression for the entropy production. It is noted that the terms $\rho_i\boldsymbol{u}_i$ are the diffusion flux vectors $\boldsymbol{J}_i$ which enter into the continuity equations (9.1.6) in the following way:

$$\rho\dot{\alpha}_i+\nabla\cdot\boldsymbol{J}_i = \rho\sigma_i \tag{9.2.5}$$

where α_i represents the mass fraction defined by

$$\alpha_i = \rho_i/\rho \tag{9.2.6}$$

and $\boldsymbol{J}_i$ is the diffusion flux vector

$$\boldsymbol{J}_i = \rho_i\boldsymbol{u}_i = \rho_i(\grave{\boldsymbol{x}}_i-\dot{\boldsymbol{x}}) \tag{9.2.7}$$

Inserting the continuity equations into (9.2.2), in which the equilibrium Gibbs relation has already been introduced, one obtains the classic expression for the entropy production (de Groot and Mazur, 1962)

$$\rho\sigma_S = -\boldsymbol{J}_S \nabla \ln \theta + \mathbf{T}^D : \nabla \dot{\boldsymbol{x}} + \rho \dot{\boldsymbol{x}} \cdot \boldsymbol{b} - \boldsymbol{J}_i \cdot \nabla(\mu_i/\theta) + \frac{\rho_i \mu_i \sigma_i}{\theta} \tag{9.2.8}$$

de Groot and Mazur (1962) refer to $\boldsymbol{b}$ as the external force instead of the body force.

What are the forces and fluxes in expression (9.2.8)? Fitts (1962) calls $\nabla \dot{\boldsymbol{x}}$ the force and $\mathbf{T}^D$ the flux. Why is it not the other way around? Consider the last term in expression (9.2.8). One usually defines the extents of reaction r_k as (de Donder, 1927)

$$d\rho_i = \rho \nu_{ik} dr_k \tag{9.2.9}$$

where ν_{ik} is the number of grams of component i produced per gram by reaction k. Then one divides (9.2.9) formally by dt and defines (Fitts, 1962)

$$\rho\sigma_i \equiv \dot{\rho}_i = \rho \nu_{ik} \dot{r}_k \tag{9.2.10}$$

However

$$\dot{\rho}_i = \rho \dot{\alpha}_i + \alpha_i \dot{\rho} \tag{9.2.11}$$

which can be written as

$$\dot{\rho}_i = \rho \dot{\alpha}_i - \rho_i \nabla \cdot \dot{\boldsymbol{x}} \tag{9.2.12}$$

using the conservation of mass (9.1.5). When (9.2.12) is substituted into (9.2.10) we do not get the continuity equation (9.2.5). Therefore only in the absence of diffusion is the chemical source σ_i equal to the time rate of change of the extents of reaction. In other words, when the divergence term is present in (9.2.2), the last term in (9.2.8) cannot be written as the bilinear sum of the affinities (de Donder, 1927)

$$A_k = -\nu_{ki} \mu_i \tag{9.2.13}$$

and their conjugate 'fluxes', namely the time rate of change of the extents of reaction. All this is to say that the pedantic formalism of classical non-equilibrium thermodynamics destroys any hoped for logical structure in non-equilibrium thermodynamics and reduces it to *ad hoc* definitions and prescriptions. According to the classical theory the fluxes are both the material fluxes as well as the time rates of change of the extents of reaction (cf. section 2.3).

Actually, there is no reason whatsoever to eliminate the divergence term in the generalised power equation (9.2.2). We form the generalised Gibbs relation on the basis that ψ must not depend on the rate variables, although it must account for all performances of work (cf. chapter 5, introduction). In the present case the constitutive relation is

$$\psi = \psi(\theta, \mathbf{F}, \alpha, \nabla \alpha) \tag{9.2.14}$$

The inclusion of the gradients of the internal state variables in the constitutive relation (9.2.14) is the starting point for our internal state variable representation of dissipative processes. In this manner, we shall be able to describe transport processes in terms of the variational equations of the internal state variables in an analogous way that we described relaxation processes in section 5.2.

Qualitatively speaking, we can justify the inclusion of the gradients of the internal state variables in the constitutive relation (9.2.14) in the following way: in order to solve the ordinary diffusion equation, we must know the boundary values of both the internal state variable and its spatial derivative. Consequently, for partial differential equations of second order in the space derivative, both the internal state variable and its spatial gradient enter into the set of state variables which define the thermodynamic process.

For any given thermokinetic process, we assume that there exists a set of kinetic equations

$$\dot{\alpha} = f(\alpha) \tag{9.2.15}$$

that govern the motion of the internal state variables. The substantial derivative of ψ can then be written as

$$\dot{\psi} = \partial_\alpha \psi f + \partial_{\nabla\alpha}\psi \cdot \nabla\dot{\alpha} + \partial_\theta \psi \dot{\theta} + \partial_{\mathbf{F}}\psi : \dot{\mathbf{F}} \tag{9.2.16}$$

Inserting (9.2.16) into the generalised power equation (9.2.2) yields

$$\partial_\alpha \psi f + \partial_{\nabla\alpha}\psi \cdot \nabla\dot{\alpha} + (\partial_{\mathbf{F}}\psi - \mathbf{S}) : \dot{\mathbf{F}} + \frac{1}{\rho}\nabla \cdot \mathbf{C}^{(i)} \cdot \boldsymbol{J}_i = -\frac{\boldsymbol{J}_S}{\rho} \cdot \nabla\theta + \theta(\pi_T - \sigma_S) \tag{9.2.17}$$

where $\mathbf{S}$ is the Piola–Kirchhoff stress tensor

$$\mathbf{S} = \frac{1}{\rho}\mathbf{T}(\mathbf{F}^T)^{-1} \tag{3.2.18}$$

Let us now consider the terms on the left-hand side of equation (9.2.17) in some detail. The constitutive equation (9.2.14) has eliminated the elastic part of the stress in the generalised power equation (9.2.2). This leaves the purely dissipative part of the stress which is a function of the rates of strain. The role of the generalised Gibbs equation (9.2.14) is to remove any non-dissipative terms on the left-hand side of (9.2.17). The kinetic equations (9.2.15) are used to establish the fact that the first two members of (9.2.17) are also purely dissipative. The kinetic equations are related to the dissipative forces $\partial_\alpha \psi$ and $\partial_{\nabla\alpha}\psi$ through the generalised equations of state. These equations connect the dissipative forces to the internal state variables and their spatial gradients. The dissipative forces are then related to the rate variables through the kinetic equations (9.2.15). If we assume that the thermodynamic process is isothermal and isochoric and choose the independent velocities so that they vanish at the non-equilibrium stationary state, then it will always be possible to write

$$-\frac{1}{\theta}\{\partial_\alpha \psi \dot{\alpha} + \partial_{\nabla\alpha}\psi \cdot \nabla\dot{\alpha}\} = R_{ij}\dot{\xi}_i\dot{\xi}_j + Q_{ij}\nabla\dot{\xi}_i \cdot \nabla\dot{\xi}_j = 2\phi \tag{9.2.18}$$

provided the thermodynamic process, characterised by the kinetic equations (9.2.15), furnishes the exactness conditions

$$R_{ij} = R_{ji} \quad \text{and} \quad Q_{ij} = Q_{ji} \tag{9.2.19}$$

These are the conditions for the existence of the dissipation function ϕ. The reduced power equation then takes the form (Lavenda, 1974)

$$\pi = \frac{1}{\rho}\nabla\cdot(\boldsymbol{J}_S - \boldsymbol{J}_D) + \dot{\eta} - 2\phi, \qquad \dot{\theta} = 0, \ \dot{\mathbf{F}} = \mathbf{0} \tag{9.2.20}$$

where we have defined the dissipative flux vector $\boldsymbol{J}_D$ as

$$\boldsymbol{J}_D = -\frac{1}{\theta}\mathbf{C}^{(i)}\boldsymbol{J}_i \tag{9.2.21}$$

Noting (9.1.9), (9.1.15) and (9.2.3), we can also write the power flux $(\boldsymbol{J}_S - \boldsymbol{J}_D)$ as

$$\theta(\boldsymbol{J}_S - \boldsymbol{J}_D) = \rho_i \varepsilon_i u_i - \mathbf{t}^{(i)}\boldsymbol{u}_i = \boldsymbol{J}_E \tag{9.2.22}$$

This permits us to write the reduced power equation (9.2.20) in the more lucid form

$$\pi - \frac{1}{\rho}\nabla\cdot\boldsymbol{J}_E = \dot{\eta} - 2\phi \tag{9.2.23}$$

This is the expression for the thermodynamic principle of the balance of power for the continua under isothermal and isochoric conditions. Expressed in words (9.2.23) states that the absorbed power less energy which flows across the surface of the system appears as the time-rate-of-change of the entropy less the energy which is dissipated in the system.

In the derivation of the thermodynamic principle (9.2.20) we have not made any concessions regarding the forms of the non-equilibrium thermodynamic potentials as is done in classical non-equilibrium thermodynamics, cf. expression (9.2.8). The only point where doubt can arise is whether there can be a stationary value of the dissipative force, since the dissipative flux (9.2.21) does not necessarily vanish in a non-equilibrium stationary state. In linear thermodynamics we know that internal motion is not permitted in a state of equilibrium (Landau and Lifshitz, 1958) and consequently there will be no stationary value of the dissipative force. For non-equilibrium stationary states this may no longer be true. In the internal state variable representation we have chosen the independent velocities so that they vanish in a non-equilibrium stationary state. Consequently, the dissipative forces also vanish there. This means that in the internal state variable representation, we are considering non-equilibrium stationary states that are completely at rest with no internal motion. The generalisation to stationary states with internal motion is straightforward: it requires the addition of a stationary value of the dissipative force in the expansion (5.2.8). The stationary state condition will then include the dissipative force and this will affect the phenomenological coefficients in the variational equations. However, the form of the variational equations will remain invariant and, for our considerations, the generalisation is purely formal.

The specific entropy will, in general, consist of two parts, cf. (5.2.22): a Gibbsian part

$$\eta' = -\partial_\theta \psi \tag{9.2.24}$$

and a non-Gibbsian part η''. The Gibbsian part η' will have the same functional dependency as ψ, while the non-Gibbsian part will be free to be a function of the rate variables if they are included in the set of state variables. Developing the

Gibbsian part of the specific entropy about a given non-equilibrium stationary state we obtain

$$\eta' = \eta'_0 + \delta\eta' - \tfrac{1}{2}(S_{ij}\xi_i\xi_j + U_{ij}\nabla\xi_i\cdot\nabla\xi_j) + (\text{higher order terms}) \quad (9.2.25)$$

Furthermore, if the variational equations of the internal state variables are expressed in terms of single degrees of freedom, then a non-Gibbsian part

$$\eta'' = -\tfrac{1}{2} M_{ij}\,\dot{\xi}_i\dot{\xi}_j \quad (9.2.26)$$

must be added to (9.2.25).

If the variational equations of the internal state variables describe a diffusion process, then there will be a 'potential energy' contribution due to elastic forces in the second variation of the specific entropy of the form $W_{ij}\nabla^2\xi_i\nabla^2\xi_j$. However, based on certain physical assumptions, it may be permissible to neglect these terms; an example will be given in section 9.5. Before this, we must derive the variational equations of the continua and show how they can be derived from a thermodynamic variational principle of the continua.

9.3 Variational equations of the continua

The reduced power equation (9.2.20) or (9.2.23) is analysed in the same manner as the power equation (5.2.14), which is the thermodynamic principle for simple materials. All terms are expanded to second order about a non-equilibrium stationary state. For the flux vectors we have the expansion

$$\boldsymbol{J}_K = \boldsymbol{J}_K^0 + \delta\boldsymbol{J}_K + \boldsymbol{j}_K \quad (9.3.1)$$

where $\delta^2\boldsymbol{J}_K \equiv \boldsymbol{j}_K$ and the subscript 'K' stands for 'E', 'S' and 'D'.

When the series expansions for the non-equilibrium thermodynamic potentials and (9.3.1) are introduced into the power equation (9.2.23) and terms of equal order are equated, we obtain the zero order power equation

$$\rho\theta\pi^0 = \nabla\cdot\boldsymbol{J}_E^0 \quad (9.3.2)$$

which determines the condition for the formation of the stationary state. It will be appreciated that condition (9.3.2) is identical to the stationarity condition that is obtained from the internal energy balance equation (9.1.12) under the conditions of constant temperature and volume. The stationarity condition (9.3.2) can also be expressed as

$$\frac{1}{\rho}\nabla\cdot\boldsymbol{J}_D^0 = \sigma_S^0 - \pi^0 \quad (9.3.3)$$

which shows that there can be internal macroscopic motion in a non-equilibrium stationary state. As we have already mentioned, it necessitates a stationary value of the dissipative force, implying that not all the independent velocities vanish in the stationary state. This is how the internal state variable representation accounts for the fact that the diffusion velocities may not vanish in a non-equilibrium stationary state.

The first order power equation furnishes the condition

$$\delta\sigma_S - \delta\pi = \frac{1}{\rho}\boldsymbol{\nabla}\cdot\delta\boldsymbol{J}_D \tag{9.3.4}$$

which reduces to the stationary force balance law (5.2.18) in the absence of the divergence of the dissipative flux vector variation. Stationary state stability is again determined by the second order or excess power equation

$$\delta^2\pi = \tfrac{1}{2}\delta^2\dot{\eta} - 2\phi + \frac{1}{\rho}\boldsymbol{\nabla}\cdot(\boldsymbol{j}_S - \boldsymbol{j}_D) \tag{9.3.5}$$

which can also be written as

$$\delta^2\pi = \delta^2\sigma_S - 2\phi - \frac{1}{\rho}\boldsymbol{\nabla}\cdot\boldsymbol{j}_D \tag{9.3.6}$$

by using the balance equation for the excess entropy

$$\tfrac{1}{2}\delta^2\dot{\eta} + \frac{1}{\rho}\boldsymbol{\nabla}\cdot\boldsymbol{j}_S = \delta^2\sigma_S \tag{9.3.7}$$

The variational equations of the continua are still determined by the variational force balance law (6.4.7). These variational equations will only describe thermodynamic processes of the continua that are both isothermal and isochoric. It is for this reason that we have called (9.2.20) and (9.2.23) reduced power equations. The problem that we are now faced with is to define the force variations in terms of the variations of the internal state variables and their spatial and temporal derivatives.

First, consider a thermodynamic process that is described by variational equations of half-degrees of freedom. The expression for the excess entropy is

$$\tfrac{1}{2}\delta^2\eta = \tfrac{1}{2}\delta^2\eta' = -\tfrac{1}{2}(S_{ij}\xi_i\xi_j + U_{ij}\nabla\xi_i\nabla\xi_j) \tag{9.3.8}$$

while the expression for the dissipation function is still given by (5.2.15). It is a general observation that in the internal state variable representation of diffusion processes, the expression for the excess entropy will contain terms that are at least one order higher in the spatial derivative than in the expression for the dissipation function.

To obtain the expression for the variation in the thermodynamic force, we observe that

$$\tfrac{1}{2}\delta^2\dot{\eta} = \tfrac{1}{2}\partial_\xi\delta^2\eta\dot{\xi} = \tfrac{1}{2}\{\partial_\xi\delta^2\eta - \boldsymbol{\nabla}\cdot(\partial_{\boldsymbol{\nabla}\xi}\delta^2\eta)\}\dot{\xi} + \tfrac{1}{2}\boldsymbol{\nabla}\cdot(\partial_{\boldsymbol{\nabla}\xi}\delta^2\eta\dot{\xi}) \tag{9.3.9}$$

On introducing the time-rate-of-change of the excess entropy (9.3.8), we get

$$\tfrac{1}{2}\delta^2\dot{\eta} = \{\boldsymbol{\nabla}\cdot(U_{ij}\boldsymbol{\nabla}\xi_j) - S_{ij}\xi_j\}\dot{\xi}_i - \boldsymbol{\nabla}\cdot(U_{ij}\boldsymbol{\nabla}\xi_j\dot{\xi}_i) \tag{9.3.10}$$

Then, comparing equations (9.3.7), (9.3.9) and (9.3.10) we identify the excess entropy production and entropy flux vector, respectively, as

$$\begin{aligned}\delta^2\sigma_S = -\chi_T\dot{\xi} &= \tfrac{1}{2}\{\partial_\xi\delta^2\eta - \boldsymbol{\nabla}\cdot(\partial_{\boldsymbol{\nabla}\xi}\delta^2\eta)\}\dot{\xi} \\ &= -\{S_{ij}\xi_j - \boldsymbol{\nabla}\cdot(U_{ij}\nabla\xi_j)\}\dot{\xi}_i\end{aligned} \tag{9.3.11}$$

$$\boldsymbol{j}_S = -\tfrac{1}{2}\rho(\partial_{\boldsymbol{\nabla}\xi}\delta^2\eta\dot{\xi}) = \rho(U_{ij}\boldsymbol{\nabla}\xi_j\dot{\xi}_i) \tag{9.3.12}$$

since the volume is constant. Finally, the expression for the thermodynamic force variations can be obtained from (9.3.11) as

$$\chi_{Ti} = -\partial_{\dot{\xi}_i}\delta^2\sigma_s = S_{ij}\xi_j - \boldsymbol{\nabla}\cdot(U_{ij}\boldsymbol{\nabla}\xi_j) \tag{9.3.13}$$

The dissipative force variations are still given by (5.2.9).

The variational equations of half-degrees of freedom are now obtained by substituting (9.3.13) and (5.2.9) into the variational force balance law (6.4.7). Assuming that the phenomenological coefficients are constants, we get

$$\chi_{Ei} = R_{ij}\dot{\xi}_j + S_{ij}\xi_j - U_{ij}\boldsymbol{\nabla}^2\xi_j, \qquad i = 1, 2, \ldots, N \tag{9.3.14}$$

With $\chi_{Ei} = 0$, the set of equations (9.3.14) are clearly recognised as the ordinary diffusion equations. In comparison with the variational equations (5.2.21) we note the presence of the diffusion term $U_{ij}\boldsymbol{\nabla}^2\xi_j$, expressing Fick's second law. The excess power equation, associated with the set of variational equations (9.3.14), is obtained by multiplying them by their conjugate velocities and summing to obtain

$$\delta^2\pi = \tfrac{1}{2}\delta^2\dot{\eta} - 2\phi + \frac{1}{\rho}\boldsymbol{\nabla}\cdot\boldsymbol{j}_S \tag{9.3.15}$$

where we have used the identity

$$\nabla^2\xi_j\dot{\xi}_i = \boldsymbol{\nabla}\cdot(\boldsymbol{\nabla}\xi_j\dot{\xi}_i) - \nabla\xi_j\cdot\boldsymbol{\nabla}\dot{\xi}_i \tag{9.3.16}$$

It is interesting to note that the divergence of the dissipative flux vector does not appear in the excess power equation (9.3.15). This implies that for diffusional processes which are described by half-degree of freedom variational equations, the stationary state conditions are formally identical to those of a simple material, namely (5.2.17) and (5.2.18).

Second, consider thermodynamic processes of the continua that are described by single degree of freedom variational equations. The specific entropy will now be the sum of a Gibbsian and a non-Gibbsian part, (9.2.25) and (9.2.26), respectively. Recall that this is due to the fact that the generalised velocities are now included in the set of state variables which was formerly comprised of the generalised displacements and their spatial derivatives. The expression for the excess entropy is

$$\tfrac{1}{2}\delta^2\eta = \tfrac{1}{2}(\delta^2\eta' + \delta^2\eta'') = -\tfrac{1}{2}(M_{ij}\dot{\xi}_i\dot{\xi}_j + S_{ij}\xi_i\xi_j + U_{ij}\nabla\xi_i\cdot\boldsymbol{\nabla}\xi_j) \tag{9.3.17}$$

while the expression for the dissipation function is given by (9.2.18). These expressions replace (9.3.8) and (5.2.15), respectively, in the case of single degree of freedom variational equations. It is to be emphasised that expression (9.3.17) is only an approximation for diffusion processes. We have neglected the 'elastic energy' contribution to the excess entropy and this can only be justified on physical grounds (cf. section 9.5).

The expression for the second variation of the entropy production is

$$\begin{aligned}\delta^2\sigma_S = -\chi_T\dot{\xi} &= \tfrac{1}{2}\{\partial_{\dot{\xi}}\delta^2\dot{\eta} - \boldsymbol{\nabla}\cdot(\partial_{\boldsymbol{\nabla}\dot{\xi}}\delta^2\dot{\eta})\}\dot{\xi} \\ &= \tfrac{1}{2}\{\partial_{\xi}\delta^2\eta + \partial_t\partial_{\dot{\xi}}\delta^2\eta - \boldsymbol{\nabla}\cdot(\partial_{\nabla\xi}\delta^2\eta)\}\dot{\xi} \\ &= -(\mathrm{M}_{ij}\ddot{\xi}_j + S_{ij}\xi_j - U_{ij}\nabla^2\xi_j)\dot{\xi}_i\end{aligned} \tag{9.3.18}$$

from which we obtain the expression for the thermodynamic force variation

$$\chi_{Ti} = -\partial_{\xi_i}\delta^2\sigma_S = (M_{ij}\ddot{\xi}_j + S_{ij}\xi_j - U_{ij}\nabla^2\xi_j) \tag{9.3.19}$$

We can also write the dissipation function (9.2.18) in the form

$$2\phi = \partial_{\dot{\xi}}\phi\dot{\xi} + \partial_{\nabla\dot{\xi}}\phi\cdot\nabla\dot{\xi} \tag{9.3.20}$$

which displays the property that the dissipation function is a second order, homogeneous function of both the generalised velocities as well as their spatial gradients. It will also be observed that expression (9.3.20) can be written as

$$2\phi = \{\partial_{\dot{\xi}}\phi - \nabla\cdot(\partial_{\nabla\dot{\xi}}\phi)\}\dot{\xi} + \nabla\cdot(\partial_{\nabla\dot{\xi}}\phi\dot{\xi}) \tag{9.3.21}$$

Then, from the definition of the dissipative force variation

$$\chi_{Di} = -\{\partial_{\dot{\xi}_i}\phi - \nabla\cdot(\partial_{\nabla\dot{\xi}}\phi)\} = -(R_{ij}\dot{\xi}_j - Q_{ij}\nabla^2\dot{\xi}_j) \tag{9.3.22}$$

equation (9.3.21) can be expressed in the form of a balance equation

$$2\phi + \frac{1}{\rho}\nabla\cdot\boldsymbol{j}_D = -\chi_D\dot{\xi} \tag{9.3.23}$$

where

$$\boldsymbol{j}_D = -\rho(\partial_{\nabla\dot{\xi}}\phi\dot{\xi}) = -\rho Q_{ij}\nabla\dot{\xi}_j\dot{\xi}_i \tag{9.3.24}$$

The balance equation (9.3.23) is comparable to the excess entropy balance equation (9.3.7) when it is written in the form

$$\tfrac{1}{2}\delta^2\dot{\eta} + \frac{1}{\rho}\nabla\cdot\boldsymbol{j}_S = -\chi_T\dot{\xi} \tag{9.3.25}$$

Subtracting (9.3.23) from (9.3.25) gives the excess power equation (9.3.5) written in the form

$$\delta^2\pi = (\chi_D - \chi_T)\dot{\xi} \tag{9.3.26}$$

The excess power equation (9.3.26) reduces to the variational force balance law (6.4.7) when we write $\delta^2\pi = -\chi_E\dot{\xi}$.

Several remarks will elucidate the significance of the foregoing results. We have constructed an internal state variable representation of transport processes of the continua that occur under isothermal and isochoric conditions. Moreover, we have chosen the independent velocities so that they vanish in the non-equilibrium stationary state. This amounts to the requirement that there be no internal macroscopic motion in the stationary state. In terms of the flux description, this means that the diffusion velocities vanish or are constant in the stationary state, since the divergence of the dissipative flux vector is what is physically observable. The stationary state will be formed by the balancing of the mechanically conservative forces only. The distinction between the cases where the dissipative force vanishes or does not vanish in the stationary state lies in the values of the phenomenological coefficients of the variational equations

$$\chi_{Ei} = M_{ij}\ddot{\xi}_j + R_{ij}\dot{\xi}_j + S_{ij}\xi_j - U_{ij}\nabla^2\dot{\xi}_j + Q_{ij}\nabla^2\xi_j, \qquad i = 1, 2, \ldots, N \tag{9.3.27}$$

which are obtained by substituting (9.3.19) and (9.3.22) into (6.4.7). The formal structure of the variational equations will be identical in the two cases.

In the internal state variable representation of thermodynamic processes of the continua, in which all the independent velocities vanish together with the dissipative forces in the stationary state, the reduced power equation (9.2.20) becomes

$$\pi = \dot{\eta} - 2\phi + \frac{1}{\rho}\nabla\cdot(\boldsymbol{J}_S - \boldsymbol{j}_D) \tag{9.3.28}$$

The reduced power equation (9.3.28) offers the following interpretation of internal dissipation: *whereas dissipation is accounted for by second order quantities in the reduced power equation, these terms have a first order effect on the stability of the motion.*

This interpretation of internal dissipation is clearly exhibited when we compare the balance equations (9.3.23) and (9.3.25). Interpretating ϕ as the time rate of change of the free energy per unit temperature and mass, it becomes apparent that $-\chi_D\dot{\xi}$ is the source of dissipation like $-\chi_T\dot{\xi}$ is the source of entropy in (9.3.25). However, (9.3.25) is a second order or excess balance equation that has been derived from the entropy balance equation (9.1.16), whereas there is no general balance equation from which we can derive (9.3.23). This implies that dissipation is a relative phenomenon which must be measured in terms of a given reference state. In the internal state variable representation, the reference state is a non-equilibrium stationary state in which there is no internal motion and, consequently, there is no internal dissipation. In contrast, the entropy production is measured from its absolute minimum value of zero at equilibrium, since in a non-equilibrium stationary state certain velocities must be maintained constant by the external constraints so as to prevent the system from relaxing back to equilibrium. Therefore internal dissipation will make its appearance in the motion about a non-equilibrium stationary state and hence it has a first order effect upon the stability of the stationary state.

9.4 Thermodynamic variational principles of the continua

In section 6.4 we showed that all thermodynamic variational principles are based, in some form or another, on the constrained principle of least dissipation of energy. In analogy with equilibrium thermodynamic potentials, we may say that the dissipation function possesses a 'conditioned' rather than a free extremum.

We recall that in the equilibrium energy representation (cf. section 1.3), the internal energy is to be minimised at constant entropy. In non-equilibrium extremum principles, the dissipation function is to be minimised subject to the appropriate thermodynamic principle that acts as the system constraint. This we found to be true in both linear and nonlinear thermodynamic variational principles. The differences between the two cases are: (1) the thermodynamic principles, and (2) the derivation of stationary state conditions from the thermodynamic variational principle in nonlinear thermodynamics. In this section we proceed to derive the thermodynamic variational principles of the continua.

The principle of least dissipation of energy is expressed in its global form as

$$\int_V \rho\phi \mathrm{d}V = \min \tag{9.4.1}$$

which is subject to the constraint imposed by the thermodynamic principle of the balance of power for the continua (9.3.28). Proceeding in the same manner as in section 6.2, we convert (9.4.1) into a free variational principle by introducing the constraint (9.3.28) explicitly into the variational expression using the method of the Lagrange undetermined multiplier. We then obtain

$$\int_V \rho\{\phi-\lambda(\pi+2\phi-\dot{\eta})\}\mathrm{d}V+\lambda\int_{\partial V}(\boldsymbol{J}_S-\boldsymbol{j}_D)\cdot\mathrm{d}\boldsymbol{S} = \min \tag{9.4.2}$$

where λ is the Lagrange undetermined multiplier. In order to determine the Lagrange multiplier, we search for the stationary value of (9.4.2) with respect to variations in the independent velocities where we can write $\dot{\boldsymbol{\alpha}} = \dot{\xi}$, since the independent velocities vanish in the stationary state. The condition of stationarity is

$$\int_V \rho\{(1-\lambda)\,(\partial_{\dot{\alpha}}\phi-\boldsymbol{\nabla}\cdot\partial_{\boldsymbol{\nabla}\dot{\alpha}}\phi)-\lambda[\partial_{\dot{\alpha}}\pi+\partial_{\dot{\alpha}}\phi-\boldsymbol{\nabla}\cdot\partial_{\boldsymbol{\nabla}\dot{\alpha}}\phi$$
$$-\ (\partial_{\dot{\alpha}}\dot{\eta}-\boldsymbol{\nabla}\cdot\partial_{\boldsymbol{\nabla}\dot{\alpha}}\dot{\eta})]\}\delta\dot{\alpha}\mathrm{d}V+\int_{\partial V}(1-\lambda)\rho\partial_{\boldsymbol{\nabla}\dot{\alpha}}\phi\delta\dot{\alpha}\cdot\mathrm{d}S = 0 \tag{9.4.3}$$

where we have used the relations

$$\partial_{\dot{\alpha}}\boldsymbol{j}_D = -\rho\partial_{\boldsymbol{\nabla}\dot{\alpha}}\phi \tag{9.4.4}$$

$$\partial_{\dot{\alpha}}\boldsymbol{J}_S = -\rho\partial_{\boldsymbol{\nabla}\dot{\alpha}}\dot{\eta} \tag{9.4.5}$$

In contrast to the entropy flux, the dissipation flux does not have a finite stationary value, since the dissipative force vanishes in the stationary state.

In order to satisfy condition (9.4.3) both the volume and surface integrals must vanish separately. This is possible only for $\lambda = 1$ which, when introduced into (9.4.2), gives the free extremum principle

$$L = \int_V \mathscr{L}\,\mathrm{d}V \equiv \int\rho(\pi+\phi-\dot{\eta})\mathrm{d}V-\int_{\partial V}(\boldsymbol{J}_S-\boldsymbol{j}_D)\cdot\mathrm{d}\boldsymbol{S} = \min \tag{9.4.6}$$

This is the free thermodynamic variational principle of the continua. In nonlinear thermodynamics the Lagrangian density $\mathscr{L}$ is developed in a series, cf. (6.4.4),

$$\mathscr{L} = \mathscr{L}^0+\mathscr{L}_1+\mathscr{L}_2+\ \ldots \tag{9.4.7}$$

where the lowest order, non-vanishing term in the series expansion determines the variational equations of the continua and hence the stability properties of the nonlinear thermodynamic process. In the same way as before, cf. section 6.4, we set $\mathscr{L}^0 = 0$ to obtain the stationary state condition

$$\pi^\circ = \frac{1}{\rho}\boldsymbol{\nabla}\cdot\boldsymbol{J}_S^0 = \sigma_S^0 \tag{9.4.8}$$

The stationarity condition is identical to (9.3.2) when we note that $\boldsymbol{J}_E^0 = \boldsymbol{\theta}\boldsymbol{J}_S^0$, since the dissipative flux vector has no influence over the stationary state condition. Moreover, condition (9.4.8) is formally identical to (5.2.17) for simple materials.

Putting $\mathscr{L}_1 = 0$ yields

$$\delta\pi = \delta\dot{\eta} + \frac{1}{\rho}\nabla\cdot\delta \boldsymbol{J}_S = \delta\sigma_S \tag{9.4.9}$$

which is formally identical to the stationary state force balance law (5.2.18). In other words, in the internal state variable representation of the continua, the stationary state conditions are formally identical to those of discrete thermodynamic processes, since the internal dissipation does not affect the conservative nature by which the non-equilibrium stationary state is maintained.

Internal dissipation does, however, influence the motion about the stationary state. This is clearly perceived when we consider the functional derivative of L_2 with respect to the independent velocities

$$\delta_{\dot{\xi}} L_2 = \int_V(\partial_{\dot{\xi}}\mathscr{L}_2 - \nabla\cdot\partial_{\nabla\dot{\xi}}\mathscr{L}_2)\delta\dot{\xi}\mathrm{d}V = 0 \tag{9.4.10}$$

since $\mathscr{L}_2$ depends on the gradients of $\dot{\xi}$. The effective Lagrangian density is

$$\mathscr{L}_2 = \rho(\delta^2\pi + \phi - \delta^2\dot{\eta}) \tag{9.4.11}$$

since the term

$$\partial_{\nabla\dot{\xi}}\mathscr{L}_2\dot{\xi} = (\boldsymbol{j}_S - \boldsymbol{j}_D) \tag{9.4.12}$$

cannot contribute to the variational equations. The condition of stationarity (9.4.10) gives the variational equations of the continua, (9.3.14) or (9.3.27), depending on the explicit expressions for the dissipation function and excess entropy. The extremum property of (9.4.1) follows from the stability requirements that both sets of eigenvalues of the matrices of the generalised friction coefficients, associated with the quadratic expression in (9.2.18), must be positive, or at most zero. This will be recalled as a necessary but not a sufficient condition of stationary state stability (cf. sections 6.1 and 7.4).

By introducing the reduced power equation directly into the constrained thermodynamic variational principle (9.4.1), we can derive several corollaries to the principle of least dissipation of energy. We recall that the principle of minimum entropy production is one corollary of the thermodynamic variational principle that is valid in linear thermodynamics (cf. section 6.2). The corresponding corollary for linear thermodynamics of the continua is

$$\int_V (\rho\sigma_S - \nabla\cdot\boldsymbol{j}_D)\mathrm{d}V = \text{min value} \tag{9.4.13}$$

It is comparable to the principle of minimum entropy production for simple materials. Moreover, if the specific entropy is constant in time then (9.4.13) can be transformed into the surface integral

$$\int_{\partial V}(\boldsymbol{j}_S - \boldsymbol{j}_D)\cdot\mathrm{d}\boldsymbol{S} = \text{min value} \tag{9.4.14}$$

which states that in the linear thermodynamics of the continua, the difference between the entropy flux and the dissipation flux tends to a minimum value.

9.5 An illustration of the internal state variable representation

As an illustration of the thermodynamic analysis of continuum processes, consider a system of chemical reactions in which the chemical intermediates are free to diffuse. Suppose that a well-defined non-equilibrium stationary state exists for a system of two chemical intermediates that participate in chemical reactions and diffusion processes. Let the internal state variables α_1 and α_2 stand for the mass fractions of the two chemical components. Then ξ_1 and ξ_2 represent the variations in the mass fractions from their stationary state values α_i^0

$$\xi_i = \alpha_i - \alpha_i^0, \qquad i = 1, 2 \tag{9.5.1}$$

The variations in the mass fractions will satisfy the variational equations that are obtained from the continuity equations (9.2.5). Suppose that the linear rates of production, σ_1 and σ_2, of the two components are given by

$$\sigma_1 = k_{11}\xi_1 + k_{12}\xi_2 \tag{9.5.2}$$

$$\sigma_2 = k_{21}\xi_1 + k_{22}\xi_2 \tag{9.5.3}$$

where the k_{ij} are the set of time independent rate constants. The variation in the diffusion flux vector is given by Fick's first law

$$\delta \boldsymbol{J}_i = -\rho D_i \boldsymbol{\nabla} \xi_i, \qquad i = 1, 2 \tag{9.5.4}$$

where the D_i are the diffusion coefficients that are assumed to be constants. Introducing expressions (9.5.2), (9.5.3) and (9.5.4) into the continuity variational equations yields the variational equations of a system composed of half-degrees of freedom

$$\dot{\xi}_i = k_{11}\xi_1 + k_{12}\xi_2 + D_1\nabla^2\xi_1 \tag{9.5.5}$$

$$\dot{\xi}_2 = k_{21}\xi_1 + k_{22}\xi_2 + D_2\nabla^2\xi_2 \tag{9.5.6}$$

Assume that the diffusion of the first component is much greater than the second. The diffusion coefficients will then be aligned such that

$$D_1 \gg D_2 \tag{9.5.7}$$

In this event, the diffusion of the second component will be of negligible importance in comparison with the diffusion of the first component. This is an example of a physical argument that can be used to justify the neglect of the 'elastic energy' contribution to the second variation of the specific entropy (9.3.17).

If one of the variations in the mass fractions is eliminated between the pair of equations (9.5.5) and (9.5.6), then the order of the resulting variational equation is raised (that is, by order one in the time derivative and order two in the space derivative). In this way variational equations of half-degrees of freedom are converted into variational equations of single degrees of freedom. In view of inequality (9.5.7), we obtain the approximate variational equation

$$\ddot{\xi} - (k_{11} + k_{22})\dot{\xi} + (k_{11}k_{22} - k_{12}k_{21})\xi + k_{22}D_1\nabla^2\xi - D_1\nabla^2\dot{\xi} = 0 \tag{9.5.8}$$

where the index on the dependent variable has been dropped since it is now superfluous. Multiplying the single degree of freedom variational equation by the generalised velocity $\dot{\xi}$ yields the excess power equation

$$2\phi - \tfrac{1}{2}\delta^2\dot{\eta} + \frac{1}{\rho}\boldsymbol{\nabla}\cdot(\boldsymbol{j}_D - \boldsymbol{j}_S) = 0 \tag{9.5.9}$$

where

$$2\phi = -(k_{11}+k_{22})\dot{\xi}^2 + D_1(\nabla\dot{\xi})^2 \tag{9.5.10}$$

$$\delta^2\eta = -\{\dot{\xi}^2 + (k_{11}k_{22} - k_{12}k_{21})\xi^2 - k_{22}D_1(\nabla\xi)^2\} \tag{9.5.11}$$

$$\boldsymbol{j}_D = -\rho D_1(\boldsymbol{\nabla}\dot{\xi}\dot{\xi}) \tag{9.5.12}$$

$$\boldsymbol{j}_S = -\rho k_{22} D_1(\boldsymbol{\nabla}\xi\dot{\xi}) \tag{9.5.13}$$

The excess power equation (9.5.9) can now be used to determine the free thermodynamic variational principle. Introducing the thermodynamic constraint into the constrained principle of least dissipation of energy (9.4.1), and determining the Lagrange undetermined multiplier by the usual procedure, we find that the second order thermodynamic Lagrangian is

$$L_2 = \int_V \mathscr{L}_2 \mathrm{d}V = \int_V \rho(\phi - \tfrac{1}{2}\delta^2\dot{\eta})\mathrm{d}V + \int_{\partial V}(\boldsymbol{j}_D - \boldsymbol{j}_S)\cdot\mathrm{d}\boldsymbol{S} = \min \tag{9.5.14}$$

The condition of stationarity is that the variational derivative of L_2 with respect to $\dot{\xi}$ vanishes, namely

$$\delta_{\dot{\xi}} L_2 = \int_V (\partial_{\dot{\xi}}\mathscr{L}_2 - \boldsymbol{\nabla}\cdot\partial_{\nabla\dot{\xi}}\mathscr{L}_2)\delta\dot{\xi}\mathrm{d}V = 0 \tag{9.5.15}$$

since $\mathscr{L}_2$ depends on the gradient of $\dot{\xi}$. The stationarity condition (9.5.15) gives the variational equation (9.5.8). Observe also that no stationary state conditions have been obtained, since we began with the variational equations (9.5.5) and (9.5.6) instead of the continuity equations.

We are now ready to compare the kinetic and thermodynamic stability criteria, using the variational equation (9.5.8) as an example. Since we are concerned with asymptotic linear stability, we take the solution of the variational equation to be of the form

$$\xi(x,t) = \Xi_1(x)\Xi_2(t) \tag{9.5.16}$$

where the spatial and temporal dependencies are

$$\Xi_1 \sim \exp(ix/\lambda) \quad \text{and} \quad \Xi_2 \sim \exp(i\omega t) \tag{9.5.17}$$

The wavelength and frequency are denoted by λ and ω, respectively. In general, these parameters will be complex. That is to say, both attenuation and damping can be exhibited by the propagating wave. However, since the two parameters are related through a dispersion equation, it is only a matter of convenience which of these parameters is chosen to be complex. If we want to analyse the waveform as a function of time, we choose the wavelength to be real. Since the frequency is complex, we know that the amplitude of the perturbation can also be complex. This requires us to express the quadratic forms of the non-equilibrium thermodynamic potentials as sums of the variables multiplied by their complex conjugates, in order that the resulting expressions be real (cf. section 7.2).

The extremum of the thermodynamic variational principle (9.5.14) is determined by the condition

$$\partial_{\dot{\xi}}\partial_{\dot{\xi}^*}\phi = -(k_{11}+k_{22}) + D_1\lambda^{-2} \geqq 0 \tag{9.5.18}$$

where ξ^* is the complex conjugate of ξ. This is the first thermodynamic criterion of stability. The second thermodynamic criterion of stability is derived from the condition that the second variation of the entropy be negative definite, or at most semidefinite (cf. theorems I and II, p. 121). Introducing the spatial dependency of the perturbation, indicated in (9.5.17), into (9.5.11) yields

$$\delta^2 \eta = -\{\dot{\xi}^2 + [(k_{11}k_{22} - k_{12}k_{21}) - k_{22}D_1\lambda^{-2}]\xi^2\} \tag{9.5.19}$$

Obviously $\delta^2 \eta$ will be negative definite or semidefinite if the condition

$$k_{11}k_{22} - k_{12}k_{21} - k_{22}D_1\lambda^{-2} \geq 0 \tag{9.5.20}$$

is satisfied. We can generalise this result by observing that any solution of the time independent variational equation is also a solution of equation (9.5.8). This permits us to cast the criterion of stability in terms of a condition that must always be satisfied by the 'potential' part of the Hamiltonian form of the second variation of the entropy. What this says is that an inequality of the form (9.5.20) must always be satisfied, independent of the 'kinetic' energy part of the excess entropy, if the system is to be stable in the Lagrange sense (cf. p. 122).

What do the thermodynamic criteria of stability, (9.5.18) and (9.5.20), mean in terms of the necessary and sufficient conditions of stability that can be derived from kinetic analysis? To answer this question we substitute (9.5.16), with the indicated functional dependencies (9.5.17), directly into the pair of variational equations (9.5.5) and (9.5.6), still working under condition (9.5.7). The algebraic set of equations will have a non-trivial solution provided

$$\omega^2 - Tr\omega + \Delta = 0 \tag{9.5.21}$$

which we recognise as the secular equation. Tr is the trace

$$Tr = k_{11} + k_{22} - D_1\lambda^{-2} \leq 0 \tag{9.5.22}$$

and Δ is the determinant

$$\Delta = k_{11}k_{22} - k_{12}k_{21} - k_{22}D_1\lambda^{-2} \geq 0 \tag{9.5.23}$$

of the coefficient matrix of the linear set of algebraic equations. Necessary and sufficient conditions of stability are given by the inequalities in (9.5.22) and (9.5.23). On comparison with the thermodynamic criteria (9.5.18) and (9.5.20), we conclude that in this case the excess power equation determines both the necessary and sufficient criteria of asymptotic, linear stability. Here, we note the curious fact that the type of thermodynamic stability criteria that can be obtained depends on the level of representation (cf. p. 109). In our illustration it is easy to see that the thermodynamic analysis will not always yield sufficient conditions of stability if the thermodynamic process is described in terms of half-degrees of freedom.

However, for a thermodynamic process composed of a single degree of freedom, the thermodynamic analysis will always yield necessary and sufficient criteria of stability. This shows why the generalised thermodynamic approach of casting all the criteria of stability into a single non-equilibrium thermodynamic potential does not allow one to obtain both the necessary and sufficient criteria of stability. Moreover the thermodynamic analysis in terms of the excess power equation works even in the presence of oscillatory phenomena, as our example has clearly illustrated. The trouble sets in when we consider coupled thermodynamic processes that are composed of single degrees of freedom. Then there is no guarantee that the symmetry of the phenomenological coefficients will be preserved by the nonlinear thermodynamic processes. In the event that the symmetry is destroyed, one either has to resort to an analysis of the complex power equation, described in chapter 7, or to a nonlinear thermodynamic field analysis, described in chapter 8.

9.6 Thermodynamic evolutionary criteria

It is well known that a fluid can be in mechanical equilibrium without being in thermal equilibrium. If the temperature is not constant throughout the system then the question arises as to the stability of such an equilibrium. It is also well known that the condition of stability of a system in mechanical equilibrium is that there be no convection (Landau and Lifshitz, 1959).

The criterion for the absence of convection is that the entropy must increase with distance. For a system bounded by a surface ∂V, the condition is

$$\nabla\eta\cdot\boldsymbol{v}=\partial_{\boldsymbol{v}}\eta>0 \qquad \text{on } \partial V \tag{9.6.1}$$

where $\boldsymbol{v}$ is the unit normal vector pointed in the outward direction on the closed surface ∂V which encloses the volume V. Condition (9.6.1) requires the directional derivative of the entropy to be positive in the absence of convection. We now ask whether there is an analogous criterion in terms of the entropy flux vector.

It has been shown that the directional derivative of the entropy flux vector can be used to determine the absence of certain classes of spatial configurations (Lavenda, 1975). In order to derive the criterion, we apply the maximum principle of partial differential equations to our internal state variable representation of thermodynamic processes of the continua. For simplicity, we treat the case of a single internal state variable but the results can easily be generalised to include several internal state variables provided the coupling between them is weak. In addition, we shall consider only the case of a linear variational equation of the continua of a half-degree of freedom. The extension to linear and nonlinear variational equations that are composed of half-degrees and single degrees of freedom can also be accomplished.

The maximum principle of partial differential equations establishes the conditions for which the system will neither display a positive maximum nor a negative minimum in the spatial profile of a dependent thermodynamic variable such as temperature, concentration, etc. For a detailed discussion on the maximum principle see Protter and Weinberger (1967). We can state the maximum principle very simply in terms of a variational equation of a half-degree of freedom that is satisfied by the variation in the internal state variable ξ, namely

$$R\dot{\xi}+S\xi-U\nabla^{2}\xi=0 \tag{9.6.2}$$

Observe that the phenomenological coefficients can also be space dependent.

Let ξ satisfy the differential inequality

$$-\chi_{T}=U\nabla^{2}\xi-S\xi\geq 0 \tag{9.6.3}$$

with $S \geqq 0$ and there are bounds on the phenomenological coefficients S and U. If ξ attains a non-negative maximum M at an interior point of V, then $\xi \equiv M$. Moreover, if ξ is continuous on ∂V and an outward directional derivative exists then (Hopf, 1952)

$$\boldsymbol{\nu}\cdot\boldsymbol{\nabla}\xi = \partial_{\boldsymbol{\nu}}\xi > 0 \qquad \text{on } \partial V \tag{9.6.4}$$

unless $\xi \equiv M$. This is indeed a logical result, since ξ must increase on leaving the system. Its maximum value can be obtained on the surface ∂V or otherwise it must be constant throughout the volume. If this was not true, then a positive maximum or a negative would occur in the volume.

The maximum principle can be used to establish a sign criterion for the excess entropy flux. From the criterion

$$\delta^2\sigma_S = -\chi_T\dot{\xi} \geq 0 \tag{9.6.5}$$

and assumption (9.6.3), we conclude that $\dot{\xi} > 0$. Expression (9.3.12) identifies the excess entropy flux in terms of the spatial and temporal derivatives of the variations in the internal state variables. Applying conditions (9.6.4) and (9.6.5) yields

$$\boldsymbol{j}_S\cdot\boldsymbol{\nu} = U\dot{\xi}\boldsymbol{\nabla}\xi\cdot\boldsymbol{\nu} \geq 0 \qquad \text{on } \partial V \tag{9.6.6}$$

The criteria derived from the maximum principle have thus been translated into a thermodynamic criterion. It states that a positive maximum or a negative minimum in the spatial distribution of the variation in an internal state variable will not occur in the system if and only if the directional derivative of the excess entropy flux is positive on the surface ∂V. Another way of expressing this criterion is

$$\partial_{\boldsymbol{\nabla}\xi}\delta^2\eta\cdot\boldsymbol{\nu} \leq 0 \qquad \text{on } \partial V \tag{9.6.7}$$

which brings out the analogy with the criterion for the absence of convection (9.6.1).

If we apply either criterion (9.6.6) or (9.6.7) to the illustration given in the last section, then we find that the condition for the absence of a positive maximum or a negative minimum in the spatial profile of the variations of the chemical components is that $k_{22} < 0$. In this event, the second variation of the entropy, (9.5.11), will be the sum of negative quadratic terms only. Conditions (9.6.6) and (9.6.7) are therefore indirectly associated with the stability of the thermodynamic process.

9.7 Thermodynamics of nonlinear dissipative wavetrains

In this section we inquire into the type of wave motion that is described by the single degree of freedom variational equations of the form (9.3.27). These equations are seen to contain both linear dispersion and dissipation. Let us consider each of these effects in some detail.

Dispersive waves are characterised by the type of solution rather than by an equation, which is in contrast to hyperbolic waves whose prototype is the ordinary wave equation. Inserting a solution of the form

$$\xi = A\cos\phi, \quad \phi = kx - \omega t \tag{9.7.1}$$

into a linear, non-dissipative wave equation, gives a dispersion equation that relates the frequency ω to the wave number k. If the phase speed $\omega(k)/k$ is constant then we have a hyperbolic equation. Alternatively, if the phase speed is not constant, but rather is a function of the wave number, then the wave equation is said to be dispersive. Although the individual waves that comprise the wavetrain still travel at the phase speed, the important quantities such as entropy propagate at the group speed $\omega'(k) \equiv d\omega(k)/dk$.

In the case of linear dispersive waves, the solution will not be given by the elementary solution (9.7.1) but rather by its Fourier integral taken over all possible wave numbers. However, if we are concerned with the asymptotic motion, then it may again be possible to describe the wave motion by the elementary solution (9.7.1) (Jeffreys and Jeffreys, 1956). The important difference between hyperbolic waves and asymptotic dispersive waves, both of whose solutions can be represented by (9.7.1), is that the amplitude A, the frequency ω, and the wave number k, are not constants but slowly varying functions of space and time. This is to say that deviations from completely periodic motion are taken into account by the slow variations in the amplitude and phase which would otherwise be parameters in the hyperbolic case that are determined by arbitrary initial and boundary conditions. If the wave equation represents a stable physical situation then it is found that the amplitude decays asymptotically as $t^{-1/2}$ (Whitham, 1974). From this result we can conclude that linear dispersion has the asymptotic effect of creating a destructive interference among the individual waves that comprise the wavetrain.

The influence of linear dissipation in the variational equations (9.3.27) is also unambiguous: it causes damping. Linear dissipation and dispersion therefore have a similar role in linear wave propagation. But what about nonlinear dissipation? Specifically consider the type of nonlinear dissipation that is found in the well-known van der Pol equation. This has the form

$$M\ddot{\xi} + \beta(\xi^2 - 1)\dot{\xi} + S\xi = 0 \tag{9.7.2}$$

We can no longer associate the parameter β with the coefficient of generalised resistance, since the system can dissipate positively as well as negatively. This depends on the sign of the coefficient of the generalised velocity. The stationary state is located at the origin of the ξ, $\dot{\xi}$-state plane. For small departures from the stationary state, the linear term is the dominant one and the system shows the tendency to depart from the stationary state. However, as the system evolves further from the stationary state, the amplitude grows and the nonlinear dissipative term is no longer negligible in comparison to the linear term. Eventually, a balance between negative linear and positive nonlinear dissipation is achieved and this produces the phenomenon of what is known in nonlinear mechanics as a 'limit cycle' (Minorsky, 1962) (cf. section 7.6). Depending on the magnitude of the parameter β, the spectrum of nonlinear oscillations can extend from the quasi-periodic case for $\beta \ll 1$ to the relaxation type of oscillations for $\beta \gg 1$.

Therefore, by releasing the restriction that the variational equations (5.2.31) must be linear, we have discovered a totally new phenomenon that could not have been predicted from the results of linear analysis. Dissipation no longer has the unique role of damping system motion; its influence on the system motion rather depends on the position of the system relative to the stationary state (that is, whether it is in the linear or nonlinear region surrounding the stationary state). Unlike linear dissipation, which must always be positive if the variational equation is to describe a real physical process, nonlinear dissipation, of the van der Pol type, may compete with linear dispersion in governing the asymptotic evolution of the nonlinear dissipative wavetrain (Lavenda, Peluso and Di Chiara, 1977). As a model system that displays both nonlinear dissipation of the van der Pol type and linear dispersion, consider the nonlinear variational equation

$$M\ddot{\xi}+\beta(\xi^2-1)\dot{\xi}+S\xi-U\xi_{,xx}=0 \tag{9.7.3}$$

where we treat for simplicity a one-dimensional wave equation propagating in the x direction.

It is a hopeless task to search for a closed form solution to equation (9.7.3). Rather, the usual procedure (Minorsky, 1962) is to try to 'fit' an elementary solution of the form (9.7.1) by letting the parameters of amplitude and phase vary so as to make the fit as close as possible to the actual, unknown solution. In order to apply this method of nonlinear mechanics, we have to make two fundamental assumptions concerning the form of wave propagation that equation (9.7.3) describes: (1) the solution is periodic, and (2) there exists a space–time scale separation. In order to comply with assumption (1), we require that $\beta \ll 1$. Assumption (2) allows us to focus our attention on the variables of amplitude and phase that describe the asymptotic evolution of the system. Needless to say, we have to consider an 'aged' system, since only then will the dispersion permit us to represent the solution in the elementary form (9.7.1).

The problem is now to derive the equations of motion that describe the long space–time evolution of the amplitude and phase. In the absence of dissipation, Whitham (1965, 1974) has succeeded in developing an extremely elegant averaged Lagrangian approach to nonlinear dispersive waves. In the presence of nonlinear dissipation, all that is necessary is to average our thermodynamic variational principle of the continua in order to derive the analogous averaged thermodynamic Lagrangian (Lavenda, Peluso and Di Chiara, 1977).

The variational principle of least dissipation of energy is expressed as

$$\int\int \rho\phi\, dx\, dt = \min \tag{9.7.4}$$

where ρ is the mass per unit length, since we are considering a one-dimensional problem. As usual, we treat (9.7.4) as a constrained thermodynamic variational principle; the constraint being the thermodynamic principle of the balance of power. In order to determine the thermodynamic principle, we multiply the nonlinear variational equation (9.7.3) by $\dot{\xi}$ to obtain

$$\dot{\eta}+\frac{1}{\rho}\boldsymbol{J}_{S,x}=2\phi \tag{9.7.5}$$

where

$$\Delta\eta=\eta-\eta_0=-\tfrac{1}{2}(M\dot{\xi}^2+S\xi^2+U\xi^2_{,x}) \tag{9.7.6}$$

$$\phi = \tfrac{1}{2}\beta\dot{\xi}^2(\xi^2 - 1) \tag{9.7.7}$$

$$J_S = \rho U \dot{\xi}\xi_{,x} \tag{9.7.8}$$

The thermodynamic constraint (9.7.5) is introduced into the variational expression (9.7.4) by the method of the Lagrange undetermined multiplier. We then obtain the free thermodynamic variational principle

$$\iint \rho(\phi - \dot{\eta})\mathrm{d}x\mathrm{d}t \qquad = \min \tag{9.7.9}$$

Since variations are to be taken between definite limits, the entropy current does not enter into the variational expression. According to our thermodynamic convention of performing all kinematically permissible variations in the velocity for a prescribed configuration of the system, it is easy to see that

$$\delta\iint \rho(\phi - \dot{\eta})\mathrm{d}x\mathrm{d}t = 0 \tag{9.7.10}$$

yields the variational equation (9.7.3), just as it does in the linear case. Our search is now for a variational principle that, when averaged, will yield the equations of motion for the amplitude and phase.

In order to derive an averaged thermodynamic Lagrangian that would be a direct generalisation of Whitham's averaged Lagrangian applied to nonlinear dispersive waves, we perform a Legendre transform on the entropy in (9.7.6) with respect to the velocity

$$\Lambda = \partial_{\dot{\xi}}\eta\dot{\xi} - \eta \tag{9.7.11}$$

since it is seen to have a non-Gibbsian part. Λ is now comparable to a mechanical Lagrangian density (cf. section 8.6). What we are actually asserting is that while the thermodynamic variational principle (9.7.9) determines the nonlinear variational equation, employing the thermodynamic variational convention, its Legendre transform (Lavenda, Peluso and Di Chiara, 1977)

$$\iint \rho(\phi - \dot{\Lambda})\mathrm{d}x\mathrm{d}t \qquad = \min \tag{9.7.12}$$

when properly averaged, will determine the equations for the amplitude and phase. This is not at all surprising when we realise that the excess entropy has a Hamiltonian form, and what we need is an actual mechanical Lagrangian density in order to determine the asymptotic equations of motion for amplitude and phase, since there is a change in the variational convention.

The averaging procedure involves a time scale separation which is implicit in the variational expression (9.7.12) due to the fact that $\beta \ll 1$. This condition on β can be introduced explicitly into (9.7.12) through a 'stretching' transformation (Whitham, 1974)

$$\partial_t \;\rightarrow\; -\omega\partial_\phi + \beta\partial_T \tag{9.7.13}$$

$$\partial_x \;\rightarrow\; k\partial_\phi + \beta\partial_X \tag{9.7.14}$$

where T and X are the independent time–space variables of the long time–space scale, namely

$$T = \beta t \qquad \text{and} \qquad X = \beta x \tag{9.7.15}$$

Now, if (9.7.13) is introduced into (9.7.12) and the result is averaged over a period of the motion, we obtain the expression for the averaged thermodynamic action

$$\overline{\mathscr{A}} = \iint \rho(\overline{\phi} - \beta\partial_T\overline{\Lambda})\mathrm{d}X\,\mathrm{d}T \qquad = \min \tag{9.7.16}$$

since

$$\partial_\phi\overline{\Lambda} \equiv \frac{1}{2\pi}\int_0^{2\pi}\partial_\phi\Lambda\,\mathrm{d}\phi = 0 \tag{9.7.17}$$

on account of the periodicity requirement. The terms in the averaged thermodynamic Lagrangian are now of the same order of magnitude, showing that the effects of dissipation are cumulative and only become noticeable over the long time–space scale of evolution.

If we are going to consider the stationary values of (9.7.16) with respect to amplitude and phase, we must be careful to retain the definitions of the frequency and wave number

$$\partial_t\phi = -\omega \qquad \text{and} \qquad \partial_x\phi = k \tag{9.7.18}$$

while giving the correct dependence of ω and k on the long time–space variables T and X. Following Whitham (1974) we write

$$\phi = \beta^{-1}\Theta(X,\ T) \tag{9.7.19}$$

and *define*

$$\partial_T\Theta \equiv -\omega(X,\ T) \qquad \text{and} \qquad \partial_X\Theta \equiv k(X,\ T) \tag{9.7.20}$$

We can now proceed to determine the stationary value of the averaged thermodynamic action (9.7.16) with respect to amplitude and phase.

The stationarity of $\overline{\mathscr{A}}$ with respect to the amplitude requires

$$\delta_A\overline{\mathscr{A}} = 0\colon\ \{\partial_A\overline{\phi} - \beta\partial_T\partial_A\overline{\Lambda}\}\delta A = 0 \tag{9.7.21}$$

which is to be evaluated over a single period of the motion. According to our assumption of periodicity, the motion behaves as if it were periodic during any single period, whereas the effects of dissipation are only noticeable over the longer space–time scale of evolution. The condition of periodicity over the short space–time scale is

$$\partial_A\overline{\Lambda} = 0\colon\ \tfrac{1}{2}(S + k^2U - M\omega^2)A^2 = 0 \tag{9.7.22}$$

which is recognised as the dispersion relation that is obtained in the absence of dissipation. This is consistent with the fact that, due to the smallness of β, dissipation acts only over the long space–time scale. Condition (9.7.22) is the condition of periodicity over the short space–time scale of the motion.

For the averaged thermodynamic action to be stationary with respect to variations in the phase, or equivalent to variations in its time derivative, requires

$$\delta_{\partial_T\Theta}\overline{\mathscr{A}} = 0\colon\ \{\partial_\omega\overline{\phi} - \beta[\partial_T\partial_\omega\overline{\Lambda} - \partial_X\partial_k\overline{\Lambda}]\}\delta\partial_T\Theta = 0 \tag{9.7.23}$$

Stationarity with respect to variations in the phase yields the equation of motion of the amplitude over the long space–time scale, namely

$$\partial_TA^2 + \omega'(k)\partial_XA^2 + A^2\omega''(k)\partial_Xk = -A^2\left(\frac{A^2}{4} - 1\right) \tag{9.7.24}$$

where the dispersion relation (9.7.22) and the consistency condition

$$\partial_T k + \partial_X \omega = 0 \tag{9.7.25}$$

obtained by equating the mixed partial derivatives of the phase, has been used in the evaluation of the stationary condition (9.7.23). The consistency condition (9.7.25) expresses the fact that there is wave conservation over the long space–time scale; it can also be written as

$$\partial_T k + \omega'(k)\partial_X k = 0 \tag{9.7.26}$$

The system of equations (9.7.24) and (9.7.26) possess a double characteristic

$$\omega'(k) = \mathrm{d}X/\mathrm{d}\,T; \qquad \mathrm{d}k/\mathrm{d}\,T = 0 \tag{9.7.27}$$

These equations can be solved on this characteristic. We find that, in the asymptotic limit, the amplitude tends to a finite stationary value (Lavenda, Peluso and Di Chiara, 1977)

$$\lim_{T \to \infty} A(T) = 2 \tag{9.7.28}$$

which is in direct contrast to the non-dissipative case where the amplitude decays as $t^{-1/2}$ due to the effect of linear dispersion. Here, we have a very interesting example of a competition between nonlinear dissipation and linear dispersion to govern the asymptotic behaviour of the nonlinear wavetrain. Furthermore, the analysis has shown that linear dispersion has a 'static' role as opposed to the 'dynamic' role played by the nonlinear dissipation in ultimately driving the system into a stationary state of wave propagation, where the amplitude is determined by the nonlinear variational equation itself. This could not have occurred for a linear equation where the amplitude and phase are determined by the arbitrary initial or boundary conditions. It is for this reason that this phenomenon has been called a 'limit wave' (Lavenda, Peluso and Di Chiara, 1977) in analogy with the limit cycle phenomenon in nonlinear mechanics.

We have gone beyond the range of nonlinear thermodynamics and applied our thermodynamic analysis to actual nonlinear irreversible processes. It will be appreciated that there is no continuity in the behaviour between linear and nonlinear irreversible processes, so that we can expect totally new and distinct phenomena to appear in the nonlinear domain about non-equilibrium stationary states. If we would venture a guess as to the future trends of non-equilibrium thermodynamics it would be directed towards the analyses of actual nonlinear irreversible processes. Such nonlinear phenomena would be much closer to the functioning of the human machine than steam engines. Our story stops here and awaits the opening of this new chapter in non-equilibrium thermodynamics.

References

Bowen, R. M. (1967). Toward a thermodynamics and mechanics of mixtures. *Archs rational Mech. Analysis,* **24,** 370–403.

de Donder, Th. (1927). *L'affinité*. Gauthier-Villars, Paris.

Eckart, C. (1940). The thermodynamics of irreversible processes I. The simple fluid. *Phys. Rev.* **58,** 267–269; The thermodynamics of irreversible processes II. Fluid mixtures. *Phys. Rev.,* **58,** 269–275.

Eringen, A. C. and Ingram, J. D. (1965). A continuum theory of chemical reacting media. I. *Int. J. engng Sci.,* **3,** 197–212.

Fitts, D. D. (1962). *Nonequilibrium Thermodynamics.* McGraw-Hill, New York,

Green, A. E. and Naghdi, P. M. (1965). A dynamical theory of interacting continua. *Int. J. engng Sci.,* **3,** 231–241.

Green, A. E. and Naghdi, P. M. (1967). A theory of mixtures. *Arch. rational Mech. Analysis,* **24,** 243–263.

de Groot, S. R. and Mazur, P. (1962). *Non-Equilibrium Thermodynamics.* North-Holland, Amsterdam.

Hopf, E. (1952). A remark on linear elliptic differential equations of the second order. *Proc. Am. math. Soc.,* **3,** 791–793.

Ingram, J. D. and Eringen, A. C. (1967). A continuum theory of chemically reacting media II. Constitutive equations of reacting fluid mixtures. *Int. J. engng Sci.,* **5,** 289–322.

Jeffreys, H. and Jeffreys, B. S. (1956). *Methods of Mathematical Physics,* 3 ed. Cambridge University Press, Cambridge.

Kelly, P. D. (1964). A reacting continuum. *Int. J. engng Sci.,* **2,** 129–153.

Landau, L. D. and Lifshitz, E. M. (1958). *Statistical Physics.* Pergamon Press, Oxford.

Landau, L. D. and Lifshitz, E. M. (1959). *Fluid Mechanics.* Pergamon Press, Oxford.

Lavenda, B. H. (1974). Principles and representations of nonequilibrium thermodynamics. *Phys. Rev. A.,* **9,** 929–943.

Lavenda, B. H. (1975). Thermodynamics of averaged motion. *Found. Phys.,* **5,** 573–589.

Lavenda, B. H., Peluso, G. and Di Chiara, A. (1977). Nonlinear dissipative and dispersive waves. *J. Inst. math. Applns,* **20,** 117–131.

Minorsky, N. (1962). *Nonlinear Oscillations.* Van Nostrand, Princeton.

Müller, I. (1968). A thermodynamic theory of mixtures of fluids. *Archs rational Mech. Analysis,* **28,** 1–39.

Protter, M. H. and Weinberger, H. F. (1967). *Maximum Principles in Differential Equations.* Prentice-Hall, Englewood Cliffs, New Jersey.

Truesdell, C. (1957). *Rend. Lincei,* **22,** 33–38, 158–166.

Truesdell, C. (1960). *Handbuch der Physik.* vol. III/1, section 255 (Ed. Flügge, S.), Springer-Verlag, Berlin.

Truesdell, C. (1969). *Rational Thermodynamics.* McGraw-Hill, New York.

Whitham, G. B. (1965). A general approach to linear and nonlinear dispersive waves using a lagrangian. *J. Fluid Mech.,* **22,** 273–283.

Whitham, G. B. (1974). *Linear and Nonlinear Waves.* Wiley-Interscience, New York.

Glossary of Principal Symbols

A 'direct', as opposed to a 'component', notation is used. Vectors and tensors are denoted by bold italic and bold roman characters, respectively. Tensors of higher order than two do not appear. The transpose of a tensor **F** is denoted by $\mathbf{F}^T$ **I** represents the unit tensor. The zero vector is denoted by $\boldsymbol{0}$, the scalar zero by 0 and the zero tensor by **0**. A sum over repeated indices is always implied. Gibbs's thermodynamic notation is adhered to, except for the absolute temperature, where t is replaced by θ. The following is a list of the most commonly used symbols. In order to follow the symbolism of rational and generalised thermodynamics as closely as possible, some duplication has been allowed for. A chapter in parentheses, following the definition of a symbol, indicates that the symbol is defined as such only in that chapter.

a_i	an internal state variable.
A_i	chemical affinity.
$\boldsymbol{A}$	thermodynamic vector potential;
$\mathscr{A}$	thermodynamic action.
$\boldsymbol{B}$	dissipative vector potential.
$\boldsymbol{b}$	specific body force.
$\mathbf{C}^{(i)}$	chemical potential tensor.
C_V	specific heat capacity at constant volume.
$\mathbf{D}$	stretching tensor.
$\mathscr{D}$	Liapounov function.
$\mathbf{E}$	coefficient matrix of the second variation of the specific internal energy.
E	internal energy of the system.
$\mathbf{F}$	deformation gradient or strain.
F	Helmholtz free energy of system.
G	Gibbs free energy of system.
$\mathscr{G}$	Onsager–Machlup function; the thermodynamic analogue of Gauss's function.
$\boldsymbol{g}$	thermal gradient, $\nabla\theta$.
$\boldsymbol{J}_i$	general symbol for a velocity flux vector.
$\boldsymbol{j}_i$	second variation of a velocity flux vector.

K_θ coefficient of thermal expansion.
$\mathbf{L}$ velocity gradient.
$\mathscr{L}$ thermodynamic Lagrangian.
$\mathbf{M}$ coefficient matrix of the second variation of the non-Gibbsian part of the specific entropy.
m_i mass of component i.
$\boldsymbol{N}$ energy flux density in velocity space.
$\boldsymbol{P}$ energy flux density in configuration space.
p hydrostatic pressure.
Q heat absorbed by system.
$\mathbf{Q}$ coefficient matrix of the specific dissipation function.
$\boldsymbol{q}$ heat flux vector.
$\mathbf{R}$ generalised resistance coefficient matrix.
r radiation energy per unit mass and time (chapter 3).
r_i extent of reaction.
$\mathbf{S}$ Piola–Kirchhoff stress tensor (chapters 3 and 9).
$\mathbf{S}$ coefficient matrix of the second variation of the Gibbsian part of the specific entropy.
S entropy of the system.
$\mathbf{T}$ symmetric stress tensor.
$\mathbf{t}^{(i)}$ local stress exerted by component i.
U velocity potential.
$\mathbf{U}$ a coefficient matrix of the second variation of the specific entropy.
u_i diffusion velocity.
V volume.
V stream function (chapter 8).
W work done on system.
X_i general symbol for the total value of a non-mechanical force.
$\dot{\boldsymbol{x}}$ barycentric velocity.
$\dot{\boldsymbol{x}}_i$ peculiar velocity.
α_i internal state variable per unit mass.
γ Coleman's internal dissipation function (chapter 3).
δ Fréchet derivative (chapter 3).
δ virtual variation.
ε specific internal energy.
ζ Legendre transform of the entropy production with respect to the velocity.
η specific entropy.
θ absolute temperature.
ι_i Onsager flux, $\dot{\alpha}_i$.
κ thermal conductivity.
λ Lagrange undetermined multiplier (chapters 6, 8 and 9).
$(\lambda+\frac{2}{3}\upsilon)$ coefficient of bulk or volume viscosity (chapter 3).
μ_i chemical potential; partial specific Gibbs free energy.
ν coefficient of shear viscosity (chapter 3).
ν_{ik} grams of component i produced per gram of reaction k.
ξ_i general symbol for a perturbation variable.
π specific power per unit temperature.

ρ mass density.

ρ_i partial mass density of component i.

σ_i general symbol for a specific source term.

ϕ specific dissipation function.

χ_i general symbol for a force variation.

ψ specific Helmholtz free energy.

ω specific work.

$\boldsymbol{\omega}$ vorticity vector.

Λ Legendre transform of the specific entropy with respect to the velocity (chapters 8 and 9).

Υ generating function.

Index

Absolute temperature 9
Additive invariance, principle of 12
Admissibility, axiom of 42
Affinity 38, 154
Anti-thermodynamic path 107
Asymptotic stability criteria 70, 111, 113, 129, 165, 173
Averaged thermodynamic action 171
  stationary conditions for 172

Balance of power, thermodynamic principle of 79-82
Barycentric velocity 151
Bifurcation point 121, 136
Bilinear covariant 125-6

Carathéodory's theorem 7
Cauchy's second law of motion 152
Characteristic equation *see* Secular equation
Characteristics 172
Chemical potential 12, 153
Chemical potential tensor 153
Circulation 126, 138
Clausius-Duhem inequality 43, 47, 56, 153
Clausius-Kelvin principle 5
Complex power method 115-19
  averaged 128-30
Composite systems 3ff., 13
Conservation of energy *see* Thermodynamics, first principle of
Conservation of mass 12, 151
Constitutive relations 45ff., 154
Continuity equations 153
Critical stability 71, 113, 120-1, *see also* Meta-stability
Curie's principle 37-9

d'Alembert's principle 99
Describing function 128
Detailed balance at equilibrium, principle of 29, 63, 97, 115, 134
Determinism, axiom of 44
Diffusion velocity 151
Dispersion relation 172
Dispersive waves 168-9
Dissipation 49-55, 81, 85, 96, 118, 146, 161, 169, 171
Dissipation function 26, 54, 64, 81, 84, 91, 94ff., 121, 144, 159
Dissipative vector potential 143
Dynamic reversibility, principle of 32, 126

Einstein formula 30
Empirical temperature 4
Energy flux 151
  in configuration space 145
  in velocity space 148
Entropy 9
  balance equation 43, 97, 152
  change in 22
  decomposition into parts 56, 86-7, 148, 156-7, 159, 171

Entropy (*continued*)
excess 60, 87, 114, 115, 117, 121, 144, 158, 159, 165
expansion of 156
flux 43, 79, 100, 152, 158, 162, 168
maximum principle 10, 19-21
representation 11-14, 82-8
substantial derivative of 81, 87
Entropy production, 28, 43, 63ff., 84, 90, 154
and dissipation function 80-1
change in 89
differential of 65
excess 69ff., 84, 117, 123, 158, 159
expansion of 65, 84
invariance of 35
Legendre transform of 68, 102
minimum principle 64-5, 99, 163
mixed 74, 117, 122
time derivative of 66
Equations of state 11, 15
generalised 68, 119, 144
Equilibrium, conditions of 13, 18
Equilibrium extremum principles 19-21
Equipresence, axiom of 44
Euclidean transformations 47
Evolution criteria 167-8
Exactness conditions *see* Integrability conditions
Extent of reaction 38, 154

Fading memory 51, 52, *see also* Materials with memory
Flow, potential 35, 37, 134-6
rotational 136-9
Fluctuations 29-34, 108, 135
Flux, diffusion 38, 153
dissipative 156ff.
material 35
thermodynamic 34ff., 97, 125
Force, balance laws 82, 85, 104, 144, 147, 162
body 43, 151, 152
circulatory 120, 122-3, 144, 145
complex 128
dissipative 82, 83, 94ff., 119, 142ff., 160
electromotive 29
external 84ff., 106, 107
generalised 81
gyroscopic 119, 122, 144, 146
motional 119, 143
positional 120, 143
thermodynamic 30, 62, 66, 71, 87, 101, 133, 142ff., 158
Fourier's law 34, 48
generalised 50
Fréchet derivative 51
Fundamental equation 11

Galilei transformations 47
Gauss' principle 64, 101-3
Generating function 91, 102
Gibbs-Duhem relation 15
non-equilibrium 67-8, 78, 80, 102
Gibbs equation 12, 13, 15, 62, 153-4
generalised 70, 77-8, 80, 155
Gibbs free energy 23
Gibbs space 13ff., 66
generalised 70, 95
Group speed 169

Hamilton's principle 99
Heat conduction 34-5, 38, 55-8, 97-101
Heat conduction inequality 49, 57
Heat flux 34, 38, 50, 97
Helmholtz free energy 10, 27
constitutive relations for 47, 51, 53, 154
minimum principle 10, 54
substantial derivative of 47, 50, 53, 54, 155
Helmholtz's reciprocal theorem 126
Hodographs 130
Hybrid state 136, *see also* Bifurcation point
Hyperbolic waves 168-9

Ideal mixtures 153
Internal dissipation inequality 52
Internal energy 4
balance equation 43, 54, 55, 97, 152
change in 22
constitutive relations for 45

Internal energy (*continued*)
minimum principle 18
representation 14-19, 79-82
Integrability conditions 6
for equilibrium potentials 11, 14
for non-equilibrium potentials 63, 68, 78, 94, 103, 133, 138, 142, 155
Irreversibility 4, 10, 22-3, 90, 108, 118

Kelvin's relations 28, 29
Kelvin's theorem 126, 139
Kinetic branches 134
Kinetic energy balance equation 151
Kinetic equations 53, 63, 112, 133, 155
linearised *see* Variational equations
Kinetic stability analysis 72, *see* Liapounov's first and second methods

Lagrangian 32, 33, 148, 171, *see also* Thermodynamic Lagrangian
Least dissipation of energy, principle of 64, 97-101, 147, 161-3, 170
Le Châtelier principle, dynamic 94-6
Le Châtelier-Braun principle, dynamic 96
Legendre transform 15-16
Liapounov function 71, 115, 127
see Liapounov's second method
Liapounov's first method 112-13, 120
Liapounov's second method 113-15
Limit cycle 129, 169
Limit wave 173
Local equilibrium, assumption of 38, 60, 62, 63, 71, 77, 153
stability criterion 66, 114
Local potential 93

Marginal stability *see* Meta-stability
Markoff process 33
Mass action, law of 134
Material frame indifference, axiom of 44, 48
Materials, simple 43, 79, 86
viscoelastic 45, 49
with memory 51-3

Maximum principles of differential equations 167
Maxwell's relations *see* Integrability conditions for equilibrium potentials
Mechanical dissipation inequaltiy 50
generalisation of 53
Mechanical power equation, complex 116
excess 85, 86
in energy representation 81
in entropy representation 84
relation to stability 120-3
separation into different orders 81-2, 84-5
Meta-stability 70, 72, 95, 113, 120-1, 136
Microscopic reversibility, principle of 30-1
Momentum balance equation 43, 151
Multistationary state transitions 134-6

Navier-Stokes fluid 52
Newtonian stress tensor 49
Nonlinear averaging techniques 127-130, *see also* Averaged thermodynamic action
Nonlinear dissipative wavetrains 168-73

Onsager-Machlup function 102
Onsager's reciprocal relations *see* Reciprocal relations
Onsager's regression hypothesis 33

Partitions, adiabatic 3
diathermal 3
Peculiar velocities 151
Pfaffian forms, properties of 5ff., 125-6
Phase, plane 134, 137
space 140, 142
speed 58, 168
transition of second order 72, 134, 135
Phenomenological coefficients, decomposition into symmetric and antisymmetric parts 34, 116, 145-6
effect on stability 120-3
significance of 34ff., 64, 95, 123-7

Phenomenological relations 31, 34, 36, 38, 63, 86,
 see also Constitutive relations *and* Variational equations
Piola-Kirchhoff stress tensor 47
Power, averaged 128
 complex 116, 118
 excess 84
 expansion of 84
 flux 156
 maximum output 89
 minimum change in 90
 total 152
Power equation 80
 generalised 153
 reduced 155, 156, 161
 separation into different orders 157-9
 see also Mechanical power equation

Quasi-static process 4
Quasi-thermodynamic stability criteria 120-3, 135, 138

Radiation energy 43, 48
Reciprocal relations 26-9, 123-7
Reciprocity theorem of Onsager 29-32
Reversibility 4, 10, 29, 89, 118

Secular equation 73, 113, 120, 122, 166
Stability with respect to diffusion 66
Stationary state, conditions for 82, 85, 135, 157, 162
 properties of 64-5, 84, 132-4
Strain tensor, rate of change of 47
Stream function 138
Stress tensor, decompostion into parts 50

Thermal conductivity 50
Thermal waves 55-8
Thermodynamic action 105
 expansion of 106-7
 minimum 106
 stationary conditions for 105, 108
Thermodynamic branch 134
Thermodynamic constraints, equilibrium 13, 20
 non-equilibrium 98, 101-3, 161
Thermodynamic Lagrangian 104ff., 147, 162
 expansion of 104, 162-3
 minimum property 105
Thermodynamic path 107
Thermodynamic processes, independent 44
Thermodynamic stability criteria, equilibrium 19
 generalised 69-74, 117, 123
Thermodynamic systems 3
Thermodynamic variational methods 98ff., 147ff., 161ff., 170ff.
Thermodynamic vector potential 143
Thermodynamics, first law of 4-5
Thermodynamics, second law of 5-11
Thermokinematics 136-9

Universal criterion of evolution 65-6

Van der Pol's equation 169
 generalisation of 170
Variables, controllable 27
 extensive 12, 29, 30
 intensive 12
 internal state 37, 53, 66, 163
 non-thermodynamic 44
 pseudo-thermodynamic 12, 29, 77, 80
 uncontrollable 27
Variational equations, derivation of 97ff., 147-8, 161-3
 half-degree of freedom 71, 74, 86, 106, 142, 158-9, 164, 167
 relation to power equation 82-8, 145, 157-61
 single degree of freedom 87, 108-9, 159-60, 164, 169, 170
Velocity potential 134-6
Viscosity, bulk, coefficient of 51
 shear, coefficient of 51
Vorticity vector 138

Waves, linear 55-8, 165
 nonlinear 168-73
Work 12, 22
 generalised 49, 77, 106, 125
 maximum principle 21-3, 88

A CATALOG OF SELECTED

DOVER BOOKS

IN SCIENCE AND MATHEMATICS

CATALOG OF DOVER BOOKS

CHALLENGING MATHEMATICAL PROBLEMS WITH ELEMENTARY SOLUTIONS, A.M. Yaglom and I.M. Yaglom. Over 170 challenging problems on probability theory, combinatorial analysis, points and lines, topology, convex polygons, many other topics. Solutions. Total of 445pp. 5⅜ × 8½. Two-vol. set.
Vol. I 65536-9 Pa. $6.95
Vol. II 65537-7 Pa. $6.95

FIFTY CHALLENGING PROBLEMS IN PROBABILITY WITH SOLUTIONS, Frederick Mosteller. Remarkable puzzlers, graded in difficulty, illustrate elementary and advanced aspects of probability. Detailed solutions. 88pp. 5⅜ × 8½.
65355-2 Pa. $3.95

EXPERIMENTS IN TOPOLOGY, Stephen Barr. Classic, lively explanation of one of the byways of mathematics. Klein bottles, Moebius strips, projective planes, map coloring, problem of the Koenigsberg bridges, much more, described with clarity and wit. 43 figures. 210pp. 5⅜ × 8½. 25933-1 Pa. $5.95

RELATIVITY IN ILLUSTRATIONS, Jacob T. Schwartz. Clear nontechnical treatment makes relativity more accessible than ever before. Over 60 drawings illustrate concepts more clearly than text alone. Only high school geometry needed. Bibliography. 128pp. 6⅛ × 9¼. 25965-X Pa. $5.95

AN INTRODUCTION TO ORDINARY DIFFERENTIAL EQUATIONS, Earl A. Coddington. A thorough and systematic first course in elementary differential equations for undergraduates in mathematics and science, with many exercises and problems (with answers). Index. 304pp. 5⅜ × 8½. 65942-9 Pa. $7.95

FOURIER SERIES AND ORTHOGONAL FUNCTIONS, Harry F. Davis. An incisive text combining theory and practical example to introduce Fourier series, orthogonal functions and applications of the Fourier method to boundary-value problems. 570 exercises. Answers and notes. 416pp. 5⅜ × 8½. 65973-9 Pa. $9.95

THE THEORY OF BRANCHING PROCESSES, Theodore E. Harris. First systematic, comprehensive treatment of branching (i.e. multiplicative) processes and their applications. Galton-Watson model, Markov branching processes, electron-photon cascade, many other topics. Rigorous proofs. Bibliography. 240pp. 5⅜ × 8½. 65952-6 Pa. $6.95

AN INTRODUCTION TO ALGEBRAIC STRUCTURES, Joseph Landin. Superb self-contained text covers "abstract algebra": sets and numbers, theory of groups, theory of rings, much more. Numerous well-chosen examples, exercises. 247pp. 5⅜ × 8½. 65940-2 Pa. $6.95

Prices subject to change without notice.

Available at your book dealer or write for free Mathematics and Science Catalog to Dept. GI, Dover Publications, Inc., 31 East 2nd St., Mineola, N.Y. 11501. Dover publishes more than 175 books each year on science, elementary and advanced mathematics, biology, music, art, literature, history, social sciences and other areas.